solutions manual
for
students

solutions manual
for
students

to accompany

Paul A. Tipler

physics

for scientists and engineers

Fourth Edition

Volume 1 Chapters 1–21

Frank J. Blatt
Professor Emeritus
Michigan State University

 W. H. FREEMAN AND COMPANY
WORTH PUBLISHERS

Solutions Manual for Students, Volume 1, Chapters 1–21
by Frank J. Blatt
to accompany
Tipler: *Physics for Scientists and Engineers,* Fourth Edition

Printed in the United States of America

ISBN: 1-57259-513-2

Printing: 1 2 3 4 5 — 03 02 01 00 99

W. H. Freeman and Company
41 Madison Avenue
New York, New York 10010
http://www.whfreeman.com

contents

Chapter **1**   Systems of Measurement   1

PART I   **mechanics**

Chapter **2**   Motion in One Dimension   5
3   Motion in Two and Three Dimensions   13
4   Newton's Laws   19
5   Applications of Newton's Laws   25
6   Work and Energy   33
7   Conservation of Energy   39
8   Systems of Particles and Conservation of Momentum   45
9   Rotation   53
10   Conservation of Angular Momentum   61
11   Gravity   67
12   Static Equilibrium and Elasticity   73
13   Fluids   79

PART II   **oscillations and waves**

Chapter **14**   Oscillations   83
15   Wave Motion   91
16   Superposition and Standing Waves   97
17   Wave–Particle Duality and Quantum Physics   103

PART III   **thermodynamics**

Chapter **18**   Temperature and the Kinetic Theory of Gases   107
19   Heat and the First Law of Thermodynamics   111
20   The Second Law of Thermodynamics   119
21   Thermal Properties and Processes   125

solutions manual
for
students

CHAPTER 1

Systems of Measurement

1* • Which of the following is *not* one of the fundamental physical quantities in the SI system?

(*a*) mass (*b*) length (*c*) force (*d*) time (*e*) All of the above are fundamental physical quantities.

(*c*) Force is *not* a fundamental physical quantity; see text.

5* • Write out the following (which are not SI units) wihout using any abbreviations. For example, 10^3 meters = 1 kilometer. (*a*) 10^{-12} boo, (*b*) 10^9 low, (*c*) 10^{-6} phone, (*d*) 10^{-18} boy, (*e*) 10^6 phone, (*f*) 10^{-9} goat

(*g*) 10^{12} bull.

(*a*) 1 picoboo (*b*) 1 gigalow (*c*) 1 microphone (*d*) 1 attoboy (*e*) 1 megaphone (*f*) 1 nanogoat (*g*) 1 terabull

---

9* • The speed of sound in air is 340 m/s. What is the speed of a supersonic plane that travels at twice the speed of sound? Give your answer in kilometers per hour and miles per hour.

1. Find the speed in m/s $v = 2(340 \text{ m/s}) = 680 \text{ m/s} = 0.680 \text{ km/s}$

2. Use [1 h/(60 min/h)(60 s/min)] = 1 $v = (0.680 \text{ km/s})/(1 \text{ h}/3600 \text{ s/h}) = 2450 \text{ km/h}$

3. Use (1 mi/1.61 km) = 1 $v = (2450 \text{ km/h})(1 \text{ mi}/1.61 \text{ km}) = 1520 \text{ mi/h}$

13* • Find the conversion factor to convert from miles per hour into kilometers per hour.

Since 1 mi = 1.61 km, (v mi/h) = (v mi/h)(1.61 km/1 mi) = $1.61v$ km/h.

17* •• A right circular cylinder has a diameter of 6.8 in and a height of 2 ft. What is the volume of the cylinder in

(*a*) cubic feet, (*b*) cubic meters, (*c*) liters?

(*a*) 1. Express the diameter in feet $D = (6.8 \text{ in})(1 \text{ ft}/12 \text{ in}) = 0.567 \text{ ft}$

    2. The volume of a cylinder is $V = (\pi D^2/4)H$ $V = [\pi(0.567 \text{ ft})^2/4](2 \text{ ft}) = 0.505 \text{ ft}^3$

(*b*) Use (1 ft/0.3048 m) = 1 $V = (0.505 \text{ ft}^3)(0.3048 \text{ m}/1 \text{ ft})^3 = 0.0143 \text{ m}^3$

(*c*) Use (1000 L/1 m₃) = 1 $V = (0.0143 \text{ m}^3)(1000 \text{ L}/1 \text{ m}^3) = 14.3 \text{ L}$

21* •• The SI unit of force, the kilogram-meter per second squared (kg·m/s²) is called the newton (N). Find the dimensions and the SI units of the constant G in Newton's law of gravitation $F = Gm_1m_2/r^2$.

1. Solve for G $G = Fr^2/m_1m_2$

2. Replace the variables by their dimensions $G = (ML/T^2)(L^2)/(M^2) = L^3/(MT^2)$

3. Use the SI units for L, M, and T Units of G are $m^3/kg \cdot s^2$

25* •• What combination of force and one other physical quantity has the dimension of power?

(ML/T^2)(dimension of Y) $= ML^2/T^3$; dimension of Y $= (ML^2/T^3)/(ML/T^2) = L/T$; Y = velocity.

29* • The prefix mega means ____. (a) 10^{-9} (b) 10^{-6} (c) 10^{-3} (d) 10^6 (e) 10^9

(d) 10^6; see Table 1-1.

33* • Express as a decimal number without using powers of 10 notation: (a) 3×10^4 (b) 6.2×10^{-3} (c) 4×10^{-6}
(d) 2.17×10^5

(a) $3 \times 10^4 = 30,000$; (b) $6.2 \times 10^{-3} = 0.0062$; (c) $4 \times 10^{-6} = 0.000004$; (d) $2.17 \times 10^5 = 217,000$.

37* • A cell membrane has a thickness of about 7 nm. How many cell membranes would it take to make a stack
1 in high?

1. The number of membranes is the total thickness $N = (1\ \text{in})/(7 \times 10^{-9}\ \text{m/membrane})$
   divided by the thickness per membrane

2. Use all SI units $N = (1\ \text{in})(2.54 \times 10^{-2}\ \text{m/1 in})/(7 \times 10^{-9}\ \text{m/membrane})$

3. Solve for N; give result to 1 significant figure $N = 4 \times 10^6$ membranes

---

41* • What are the advantages and disadvantages of using the length of your arm for a standard length?

The advantage is that the length measure is always with you. The disadvantage is that arm lengths are not uni-

form, so if you wish to purchase a board of "two arm lengths" it may be longer or shorter than you wish, or else

you may have to physically go to the lumber yard to use your own arm as a measure of length.

45* • If one could count $1 per second, how many years would it take to count 1 billion dollars (1 billion $= 10^9$)?

It would take 10^9 seconds or $(10^9\ \text{s})(1\ \text{h}/3600\ \text{s})(1\ \text{day}/24\ \text{h})(1\ \text{y}/365\ \text{days}) = (10^9\ \text{s})(1\ \text{y}/3.154 \times 10^7\ \text{s}) = 31.7$ y.

49* •• The angle subtended by the moon's diameter at a point on the earth is about 0.524°. Use this and the fact that

the moon is about 384 Mm away to find the diameter of the moon. (The angle subtended by the moon θ is

approximately D/r_m, where D is the diameter of the moon and r_m is the distance to the moon.)

Note that $\theta \approx D/r_m$ for small values of θ only if θ is expressed in radians, and that radians are dimensionless.

1. Find θ in radians   (0.524 deg)$(2\pi\ \text{rad}/360\ \text{deg}) = 0.00915$ rad

2. Use $\theta = D/r_m$ and solve for D $D = \theta r_m = (0.00915)(384\ \text{Mm}) = 3.51$ Mm

---

53* •• Evaluate the following expressions. (a) $(5.6 \times 10^{-5})(0.0000075)/(2.4 \times 10^{-12})$;

(b) $(14.2)(6.4 \times 10^7)(8.2 \times 10^{-9}) - 4.06$; (c) $(6.1 \times 10^{-6})^2(3.6 \times 10^4)^3/(3.6 \times 10^{-11})^{1/2}$;

(d) $(0.000064)^{1/3}/[(12.8 \times 10^{-3})(490 \times 10^{-1})^{1/2}]$.

(a) $(5.6 \times 10^{-5})(7.5 \times 10^{-6})/(2.4 \times 10^{-12}) = 1.8 \times 10^2$ to two significant figures.

(b) $(14.2)(6.4 \times 10^7)(8.2 \times 10^{-9}) - 4.06 = 7.45 - 4.06 = 3.4$ to two significant figures.

(c) $(6.1 \times 10^{-6})^2(3.6 \times 10^4)^3/(3.6 \times 10^{-11})^{1/2} = 2.9 \times 10^8$ to two significant figures.

(d) $(6.4 \times 10^{-5})^{1/3}/[(12.8 \times 10^{-3})(49.0)^{1/2}] = 0.45$ to two significant figures.

57* •• Beer and soft drinks are sold in aluminum cans. The mass of a typical can is about 0.018 kg. (*a*) Estimate the number of aluminum cans used in the United States in one year. (*b*) Estimate the total mass of aluminum in a year's consumption from these cans. (*c*) If aluminum returns $1/kg at a recycling center, how much is a year's accumulation of aluminum cans worth?

(*a*) The population of the U.S. is about 3×10^8 persons. Assume 1 can per person per day. In one year the total number of cans used is $(3 \times 10^8 \text{ persons})(1 \text{ can/person-day})(365 \text{ days/y}) = 1 \times 10^{11}$ cans/y.

(*b*) Total mass of aluminum per year = $(1 \times 10^{11} \text{ cans/y})(1.8 \times 10^{-2} \text{ kg/can}) = 2 \times 10^9$ kg/y.

(*c*) At $1/kg this amounts to $2 billion.

61* ••• The period T of a simple pendulum depends on the length L of the pendulum and the acceleration of gravity g (dimensions L/T²). (*a*) Find a simple combination of L and g which has the dimensions of time. (*b*) Check the dependence of the period T on the length L by measuring the period (time for a complete swing back and forth) of a pendulum for two different values of L. (*c*) The correct formula relating T to L and g involves a constant which is a multiple of π, and cannot be obtained by the dimensional analysis of part (*a*). It can be found by experiment as in (*b*) if g is known. Using the value $g = 9.81$ m/s² and your experimental results from (*b*), find the formula relating T to L and g.

(*a*) 1. Write $T = CL^a g^b$ and express dimensionally $\qquad T = L^a (L/T^2)^b = L^{a+b} T^{-2b}$.

     2. Solve for a and b $\qquad\qquad\qquad\qquad\qquad -2b = 1, b = -\tfrac{1}{2}; a + b = 0, a = \tfrac{1}{2}$.

     3. Write the expression for T $\qquad\qquad\qquad T = C\sqrt{L/g}$

(*b*) Check by using pendulums of lengths 1 m and 0.5 m; the periods should be about 2 s and 1.4 s.

(*c*) Using $L = 1$ m, $T = 2$ s, $g = 9.81$ m/s², $C = (2.0 \text{ s})/\sqrt{1.0 \text{ m}/9.81 \text{ m/s}^2} = 6.26 = 2\pi$. $T = 2\pi\sqrt{L/g}$.

CHAPTER 2

Motion in One Dimension

1* • What is the approximate average velocity of the race cars during the Indianapolis 500?

Since the cars go around a closed circuit and return nearly to the starting point, the displacement is nearly zero, and the average velocity is zero.

5* • (*a*) An electron in a television tube travels the 16-cm distance from the grid to the screen at an average speed of 4×10^7 m/s. How long does the trip take? (*b*) An electron in a current-carrying wire travels at an average speed of 4×10^{-5} m/s. How long does it take to travel 16 cm?

(*a*) From Equ. 2-3, $\Delta t = \Delta s/$(av. speed) $\Delta t = (0.16 \text{ m})/(4 \times 10^7 \text{ m/s}) = 4 \times 10^{-9} \text{ s} = 4 \text{ ns}$

(*b*) Repeat as in (*a*) $\Delta t = (0.16 \text{ m})/(4 \times 10^{-5} \text{ m/s}) = 4 \times 10^3 \text{ s} = 4 \text{ ks}$

9* • As you drive down a desert highway at night, an alien spacecraft passes overhead, causing malfunctions in your speedometer, wristwatch, and short-term memory. When you return to your senses, you can't tell where you are, where you are going, or even how fast you are traveling. The passenger sleeping next to you never woke up during this incident. Although your pulse is racing, hers is steady at 55 beats per minute. (*a*) If she has 45 beats between the mile markers posted along the road, determine your speed. (*b*) If you want to travel at 120 km/h, how many heartbeats should there be between mile markers?

(*a*) Find the time between mile markers $\Delta t = (45 \text{ beats/mile})/(55 \text{ beats/min}) = 0.818 \text{ min}$

     $v = \Delta s/\Delta t$ $v = 1 \text{ mi}/0.818 \text{ min} = 1.22 \text{ mi/min} = 73.3 \text{ mi/h}$

(*b*) N = (beats/min)(60 min/h)/(v mi/h)   N = $(55 \times 60 \text{ beats/h})/[(120 \text{ km/h})/(1.61 \text{ km/mi})] = 44.3$

13* •• John can run 6.0 m/s. Marcia can run 15% faster than John. (*a*) By what distance does Marcia beat John in a 100-m race? (*b*) By what time does Marcia beat John in a 100-m race?

(*a*) 1. Find the running speed for Marcia $v_M = 1.15(6.0 \text{ m/s}) = 6.9 \text{ m/s}$

     2. Find the time for Marcia $t_M = (100 \text{ m})/6.9 \text{ m/s}) = 14.5 \text{ s}$

     3. Find the distance covered by John in 14.5 s $s_J = (6.0 \text{ m/s})(14.5 \text{ s}) = 87 \text{ m}$; distance = 13 m

(*b*) Find the time required by John $t_J = (100 \text{ m})/(6 \text{ m/s}) = 16.7 \text{ s}$; time difference = 2.2 s

---

17* • If the instantaneous velocity does not change, will the average velocities for different intervals differ?

No, they will not. For constant velocity, the instantaneous and average velocities are equal.

21* •   Using the graph of x versus t in Figure 2-25, (*a*) find the average velocity between the times $t = 0$ and $t = 2$ s.
(*b*) Find the instantaneous velocity at $t = 2$ s by measuring the slope of the tangent line indicated.

(*a*) Find Δx from graph; $v_{av} = \Delta x/\Delta t$ $\Delta x = 2$ m, $\Delta t = 2$ s; $v_{av} = 1$ m/s

(*b*) From graph, tangent passes through points   Slope of tangent line is (4 m)/(2 s) = 2 m/s

    $x = 0$, $t = 1$ s; $x = 4$ m, $t = 3$ s. $v(t = 2$ s) = 2 m/s

---

25* ••   The position of a body oscillating on a spring is given by $x = A \sin \omega t$, where A and ω are constants with
values $A = 5$ cm and $\omega = 0.175$ s^{-1}. (*a*) Sketch x versus t for $0 \le t \le 36$ s. (*b*) Measure the slope of your graph at
$t = 0$ to find the velocity at this time. (*c*) Calculate the average velocity for a series of intervals beginning at $t = 0$
and ending at $t = 6, 3, 2, 1, 0.5$, and 0.25 s. (*d*) Compute dx/dt and find the velocity at time $t = 0$.

(*a*) The plot of x versus t is shown

(*b*) The slope of the dotted line (tangent at $t = 0$)
    is 0.875; $v(0) = 0.875$ cm/s

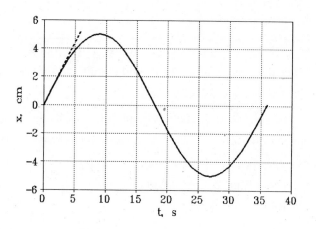

(*c*)

t	x	$\Delta x/\Delta t$
6	4.34	0.723
3	2.51	0.835
2	1.71	0.857
1	0.174	0.871
0.5	0.437	0.874
0.25	0.219	0.875

(*d*) $dx/dt = A\omega \cos \omega t$; at $t = 0 \cos \omega t = 1$;
    dx/dt at $t = 0$ is $A\omega = 0.875$ cm/s

29* ••   Margaret has just enough gas in her speedboat to get to the marina, an upstream journey that takes 4.0 hours.
Finding it closed for the season, she spends the next 8.0 hours floating back downstream to her shack. The entire
trip took 12.0 h; how long would it have taken if she had bought gas at the marina?

Let D = distance to marina, v_W = velocity of stream, v_{rel} = velocity of boat under power relative to stream.

1. Express the times of travel <u>with gas</u> in terms of $t_1 = D/(v_{rel} - v_W) = 4$ h
    D, v_W, and v_{rel} $t_2 = D/(v_{rel} + v_W)$

2. Express the time required to drift distance D $t_3 = D/v_W = 8$ h; $v_W = D/(8$ h)

3. From $t_1 = 4$ h, find v_{rel} $4(v_{rel} - D/8) = D$; $v_{rel} = 1.5D/(4$ h)

4. Solve for t_2 $t_2 = D/[(1.5D/4$ h) $+ (D/8$ h)] $= 2$ h

5. Add t_1 and t_2 $t_{tot} = t_1 + t_2 = 6$ h

33* •   Is it possible for a body to have zero velocity and nonzero acceleration?

Yes, it is. An object tossed up has a constant downward acceleration; at its maximum height, its instantaneous
velocity is zero.

37* •   A BMW M3 sports car can accelerate in third gear from 48.3 km/h (30 mi/h) to 80.5 km/h (50 mi/h) in 3.7 s. (*a*) What is the average acceleration of this car in m/s^2? (*b*) If the car continued at this acceleration for another second, how fast would it be moving?

(*a*) 1. Use Equ. 2-8

    2. Convert to m/s^2

(*b*) In 1 s, its speed increases by 8.7 km/h

$a_{av} = [(80.5 - 48.3) \text{ km/h}]/(3.7 \text{ s}) = 8.7 \text{ km/h·s}$

$(8.7 \times 10^3 \text{ m/h·s})(1 \text{ h}/3600 \text{ s}) = 2.42 \text{ m/s}^2$

$v = (80.5 + 8.7) \text{ km/h} = 89.2 \text{ km/h}$

41* •   Identical twin brothers standing on a bridge each throw a rock straight down into the water below. They throw rocks at exactly the same time, but one hits the water before the other. How can this occur if the rocks have the same acceleration?

The initial downward velocities of the two rocks are not the same.

45* •   An object projected up with initial velocity *v* attains a height *H*. Another object projected up with initial velocity 2*v* will attain a height of (*a*) 4*H*, (*b*) 3*H*, (*c*) 2*H*, (*d*) *H*.

(*a*) 4*H*; from Equ. 2-15, with $a = -g$ and $v = 0$ at top of trajectory, $H = v_0^2/2g$, v_0 is initial velocity. So $H \propto v_0^2$.

49* ••   An object is dropped from rest. If the time during which it falls is doubled, the distance it falls will (*a*) double, (*b*) decrease by one-half, (*c*) increase by a factor of four, (*d*) decrease by a factor of four, (*e*) remain the same.

(*c*) increase by a factor of 4; see Equ. 2-14: $\Delta x \propto t^2$ if $v_0 = 0$.

53* •   An object with constant acceleration has velocity $v = 10$ m/s when it is at $x = 6$ m and $v = 15$ m/s when it is at $x = 10$ m. What is its acceleration?

Use Equ. 2-15

$a = [(15^2 - 10^2) \text{ m}^2/\text{s}^2]/[2(4 \text{ m})] = 15.6 \text{ m/s}^2$

57* ••   A ball is thrown upward with an initial velocity of 20 m/s. (*a*) How long is the ball in the air? (*b*) What is the greatest height reached by the ball? (*c*) When is the ball 15 m above the ground?

(*a*) 1. Take upward as positive; use Equ. 2-14

    2. Solve for *t*

(*b*) $H = v_0^2/2g$

(*c*) 1. Use Equ. 2-14

    2. Use quadratic formula $t = \dfrac{-b \pm \sqrt{b^2 - 4ac}}{2a}$

$\Delta x = 0 = (20 \text{ m/s})t - \frac{1}{2}(9.81 \text{ m/s}^2)t^2$

$t = 0; t = 4.08 \text{ s}; t = 4.08 \text{ s}$ is the proper result

$H = (400 \text{ m}^2/\text{s}^2)/[2(9.81 \text{ m/s}^2)] = 20.4 \text{ m}$

$15 \text{ m} = (20 \text{ m/s})t - \frac{1}{2}(9.81 \text{ m/s}^2)t^2$

$t = 0.991 \text{ s}, t = 3.09 \text{ s}$; both are acceptable solutions

---

61* ••   An automobile accelerates from rest at 2 m/s^2 for 20 s. The speed is then held constant for 20 s, after which there is an acceleration of −3 m/s^2 until the automobile stops. What is the total distance traveled?

1. Determine the distance traveled during first 20 s and the speed at the end of first 20 s

2. Find Δx_2 = distance covered 20 s $\le t \le$ 40 s

3. Find Δx_3 = distance during deceleration

4. Find total distance

$v(20) = at = (2 \text{ m/s}^2)(20 \text{ s}) = 40 \text{ m/s}$

$\Delta x_1 = v_{av}t = (20 \text{ m/s})(20 \text{ s}) = 400 \text{ m}$

$\Delta x_2 = (40 \text{ m/s})(20 \text{ s}) = 800 \text{ m}$

$\Delta x_3 = (40 \text{ m/s})^2/[2(3 \text{ m/s}^2)] = 267 \text{ m}$

$x = \Delta x_1 + \Delta x_2 + \Delta x_3 = 1467 \text{ m}$

65* ••   To win publicity for her new CD release, Sharika, the punk queen, jumps out of an airplane without a parachute. She expects a stack of loose hay to break her fall. If she reaches a speed of 120 km/h prior to impact, and if a 35 g deceleration is the greatest deceleration she can withstand, how high must the stack of hay be in order for her to survive? Assume uniform acceleration while she is in contact with the hay.

1. Use Equ. 2-15

$$H = \frac{(1.20 \times 10^5 \text{ m/h})^2}{(3600 \text{ s/1 h})^2 [2 \times 35 \times (9.81 \text{ m/s}^2)]} = 1.62 \text{ m}$$

69* •   A stone is thrown vertically from a cliff 200 m tall. During the last half second of its flight the stone travels a distance of 45 m. Find the initial velocity of the stone.

We take <u>down</u> as the positive direction. Let v_1 be the velocity ½ s before impact, v_f velocity at impact.

1. Find v_{av} during last ½ s $v_{av} = \frac{1}{2}(v_1 + v_f) = (45/0.5)$ m/s; $v_1 + v_f = 180$ m/s

2. Write v_f in terms of v_1 and g $v_f = v_1 + gt; \; v_f = v_1 + (9.81 \times 0.5)$ m/s

3. Solve for v_f $v_f = 92.5$ m/s

4. Use Equ. 2-15 to find v_0 $v_0 = \sqrt{92.5^2 - 2 \times 9.81 \times 200}$ m/s $= \pm 68$ m/s; the stone
                                                 may be thrown either up or down

---

73* ••   A rocket is fired vertically with an upward acceleration of 20 m/s². After 25 s, the engine shuts off and the rocket continues as a free particle until it reaches the ground. Calculate (a) the highest point the rocket reaches, (b) the total time the rocket is in the air, (c) the speed of the rocket just before it hits the ground.

Take up as the positive direction.

(a) 1. Find x_1 and v_1 at $t = 25$ s; use Equ. 2-14 $x_1 = \frac{1}{2}(20)(25)^2$ m $= 6250$ m; $v_1 = 20 \times 25$ m/s $= 500$ m/s

   2. Find $x_2 =$ distance above x_1 when $v = 0$ $x_2 = (500)^2/(2 \times 9.81)$ m $= 1.274 \times 10^4$ m

   3. Total height $= x_1 + x_2$ $H = 1.90 \times 10^4$ m $= 19.0$ km

(b) 1. Find time, t_2, for part (a) $t_2 = x_2/v_{av} = (1.274 \times 10^4/250)$ s $= 51$ s

   2. Find time, t_3, to drop 19.0 km $t_3 = [2(1.90 \times 10^4)/9.81]^{\frac{1}{2}}$ s $= 62.2$ s

   3. Total time, $T = 25$ s $+ t_2 + t_3$ $T = (25 + 51 + 62.2)$ s $= 138$ s $= 2$ min 18 s

(c) Use Equ. 2-12 $v_f = (9.81)(62.2)$ m/s $= 610$ m/s

77* ••   A rock dropped from a cliff falls one-third of its total distance to the ground in the last second of its fall. How high is the cliff?

1. Write the final speed in terms of H and g $v_f^2 = 2gH$

2. Write the average speed in the last second $v_{av} = \frac{1}{2}[v_f + (v_f - g)] = (v_f - g/2)$

3. Set the distance in last second $= H/3$; solve for v_f $(v_f - g/2) = H/3; \; v_f = H/3 + g/2$

4. Set $v_f^2 = 2gH$ and solve for H $2gH = H^2/9 + gH/3 + g^2/4; \; H = 14.85g = 145.7$ m

---

81* ••   A motorcycle policeman hidden at an intersection observes a car that ignores a stop sign, crosses the intersection, and continues on at constant speed. The policeman takes off in pursuit 2.0 s after the car has passed the stop sign, accelerates at 6.2 m/s² until his speed is 110 km/h, and then continues at this speed until he catches the car. At that instant, the car is 1.4 km from the intersection. How fast was the car traveling?

1. Find the time of travel of policeman: t_1 is the time $t_1 = (1.10 \times 10^3$ m/h)/[(3600 s/1 h)(6.2 m/s²)] $= 4.93$ s
   of acceleration, t_2 the time of travel at 110 km/h; $d_1 = \frac{1}{2}(30.5$ m/s)(4.93 s) $= 75.3$ m; $d_2 = (1400 - 75.3)$ m
   and d_1 and d_2 the corresponding distances $= 1325$ m; $t_2 = d_2/(30.6$ m/s) $= 43.3$ s

2. The time of travel of car is 2.0 s + t_1 + t_2 $t_C = (2.0 + 43.3 + 4.93)$ s = 50.2 s

3. Find the speed of the car $v_C = (1400/50.2)$ m/s = 27.9 m/s = 100.4 km/h

---

85* •• A train pulls away from a station with a constant acceleration of 0.4 m/s². A passenger arrives at the track 6.0 s after the end of the train has passed the very same point. What is the slowest constant speed at which she can run and catch the train? Sketch curves for the motion of passenger and the train as functions of time.

1. The critical conditions are $v_T = v_P$ and $x_T = x_P$ $v_T = 0.4t$ m/s; $x_T = 0.2t^2$ m; $x_P = v_P(t - 6\ \text{s})$

                                                     $0.2t^2 = 0.4t(t - 6)$

2. Solve for t $0.2t^2 = 2.4t$; $t = 12$ s

3. Find v_T at $t = 12$ s, which is also v_{PC} $v_{PC} = 4.8$ m/s

The positions of the train and passenger as functions of time are shown in the adjoining figure.

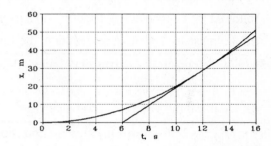

89* ••• The Sprint missile, designed to destroy incoming ballistic missiles, can accelerate at 100g. If an ICBM is detected at an altitude of 100 km moving straight down at a constant speed of 3 × 10⁴ km/h and the Sprint missile is launched to intercept it, at what time and altitude will the interception take place? (*Note:* You can neglect the acceleration due to gravity in this problem. Why?)

1. Neglect g; see below. Find x_{ICBM}, x_S $x_S = \frac{1}{2}at^2$; $x_{ICBM} = H - vt$

2. Express v in m/s $v = (3 \times 10^7$ m/h$)(1$ h/3600 s$) = 8.33 \times 10^3$ m/s

3. Set $x_{ICBM} = x_S$ with $H = 10^5$ m and find t $\frac{1}{2} \times 981t^2 + 8.33 \times 10^3t - 10^5 = 0$; $t = 8.12$ s

4. Find x_{ICBM} $x_{ICBM} = [981 \times 8.12^2/2]$ m = 3.24 × 10⁴ m = 32.4 km

Note: In 8 s, the change in velocity Δv of the ICBM due to g is less than 80 m/s , i.e., less than 1% of v; also, g = 1% of a_S. So the result is good to about 1%. Also, if $a_S = 100g$ is taken to be the possible horizontal acceleration, then both objects suffer the same downward acceleration due to gravity, and this contribution cancels.

93* •• The velocity of a particle is given by $v = 7t^2 - 5$, where t is in seconds and v is in meters per second. Find the general position function $x(t)$.

$$x(t) = \int (7t^2 - 5)\,dt = (7/3)t^3 - 5t + C.$$

97* ••• Figure 2-33 shows a plot of x versus t for a body moving along a straight line. Sketch rough graphs of v versus t and a versus t for this motion.

Note: The curve of x versus t appears to be a sine curve; hence $v(t) = dx/dt$ is a cosine curve and $a(t) = dv/dt$ is a negative sine curve.

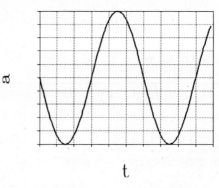

101* •• On a graph showing position on the vertical axis and time on the horizontal axis, a parabola that opens upward represents (*a*) a positive acceleration. (*b*) a negative acceleration. (*c*) no acceleration. (*d*) a positive followed by a negative acceleration. (*e*) a negative followed by a positive acceleration.

(*a*) it represents a positive acceleration; the slope—velocity—is increasing.

105* •• Which graph of *v* versus *t* in Figure 2-34 best describes the motion of a particle with negative velocity and negative acceleration?

(*d*) $v \leq 0$ and the slope of $v(t)$ is negative.

109* •• Figure 2-36 shows nine graphs of position, velocity, and acceleration for objects in linear motion. Indicate the graphs that meet the following conditions: (*a*) Velocity is constant (*b*) Velocity has reversed its direction (*c*) Acceleration is constant (*d*) Acceleration is not constant. Which graphs of velocity and acceleration are mutually consistent?

(*a*) *a, f, i*; (*b*) *c, d*; (*c*) *a, d, e, f, h, i*; (*d*) *b, c, g*. The graphs *d* and *h*, and *f* and *i* are mutually consistent.

113* •• The cheetah can run as fast as $v_1 = 100$ km/h, the falcon can fly as fast as $v_2 = 250$ km/h, and the sailfish can swim as fast as $v_3 = 120$ km/h. The three of them run a relay with each covering a distance *L* at maximum speed. What is the average speed *v* of this triathlon team?

1. Express the time required for each animal $t_1 = L/v_1;\ t_2 = L/v_2;\ t_3 = L/v_3$

2. Write the total time, Δt $\Delta t = L(1/v_1 + 1/v_2 + 1/v_3)$

3. Find $v = \Delta x / \Delta t = 3L/\Delta t$; use values for v_1 etc. $v = [3/(0.01 + 0.004 + 0.00833)]$ m/s = 134 m/s

117* •• Consider the velocity graph in Figure 2-38. Assuming $x = 0$ at $t = 0$, write correct algebraic expressions for $x(t)$, $v(t)$, and $a(t)$ with appropriate numerical values inserted for all constants.

1. Write $v(t)$; note that *a* is constant and < 0 $v(t) = (50 - 10t)$ m/s; $a = -10$ m/s^2

2. $x(t) = \int v(t)dt$ $x(t) = 50t - 5t^2$

121* •• Without telling Sally, Joe made travel arrangements that include a stopover in Toronto to visit Joe's old buddy. Sally doesn't like Joe's buddy and wants to change their tickets. She hops on a courtesy motor scooter and begins accelerating at 0.9 m/s^2 toward the ticket counter to make arrangements. As she begins moving, Joe is 40 m behind her, running at constant speed of 9 m/s. (*a*) How long does it take for Joe to catch up with her? (*b*) What is the time interval during which Joe remains ahead of Sally?

(*a*) 1. Write expressions for x_S and x_J $x_S = 0.45t^2$ m; $x_J = (-40 + 9t)$ m

     2. Set $x_S = x_J$ to obtain equation for *t* $0.45t^2 - 9t + 40 = 0$

3. Solve for t; keep smallest answer $t = 6.67$ s, $t = 13.33$ s; $t = 6.67$ s

(b) Sally will catch Joe at 13.33 s $\Delta t = 6.67$ s

125* ••   Repeat Problem 124, but with the runner attempting to steal third base, starting from second base with a lead of 3 m.

We neglect the time of acceleration of the runner; we will assume that the speed of a fast ball is 28 m/s and use that also for the speed of the ball thrown by the catcher. We will take 0.6 s as the reaction time of the catcher. The time taken by the runner is (23/9.5) s = 2.42 s. The time of flight of the ball is (18.5 + 26)/28 s = 1.59 s. Add to that the reaction time of 0.6 s to obtain a total time for the ball to reach third base of 2.19 s. The ball will reach third base about one quarter second before the runner. A good umpire will call him out!

129* •••   Suppose that a particle moves in a straight line such that, at any time t, its position, velocity, and acceleration all have the same numerical value. Give the position x as a function of time.

We are given that $v = 1 \times x = 1 \times a$. So $dx/dt = x$, and integrating we obtain $t - t_0 = \ln(x/x_0)$ or $x(t) = x_0 e^{t-t_0}$. The velocity and acceleration are obtained from $v = dx/dt$ and $a = dv/dt$: $v(t) = x_0 e^{t-t_0} = a(t)$.

CHAPTER 3

Motion in Two and Three Dimensions

1* • Can the magnitude of the displacement of a particle be less than the distance traveled by the particle along its path? Can its magnitude be more than the distance traveled? Explain.

The magnitude of the displacement can be less but never more than the distance traveled. If the path is a semi-circle of radius R, the distance traveled is $2\pi R$, the magnitude of the displacement is $2R$. The maximum magnitude of the displacement occurs when the path is a straight line; then the two quantities are equal.

5* • (a) A man walks along a circular arc from the position $x = 5$ m, $y = 0$ to a final position $x = 0$, $y = 5$ m. What is his displacement? (b) A second man walks from the same initial position along the x axis to the origin and then along the y axis to $y = 5$ m and $x = 0$. What is his displacement?

(a), (b) Since the initial and final positions are the same, the displacements are equal. $D = (-5\,i + 5\,j)$ m.

9* • Can a component of a vector have a magnitude greater than the magnitude of the vector? Under what circumstances can a component of a vector have a magnitude equal to the magnitude of the vector?

No; $A_x = A \cos\theta \leq A$. $A_x = A$ for $\theta = 0$, i.e., if the vector is along the component direction.

13* • A velocity vector has an x component of $+5.5$ m/s and a y component of -3.5 m/s. Which diagram in Figure 3-32 gives the direction of the vector?

(b) The vector is in the fourth quadrant; $\theta = \tan^{-1}(-3.5/5.5) = -32.5°$.

17* • Find the magnitude and direction of the following vectors: (a) $A = 5\,i + 3\,j$; (b) $B = 10\,i - 7\,j$; (c) $C = -2\,i - 3\,j + 4\,k$.

(a), and (b) Use Equs. 3-4 and 3-5 $A = 5.83$, $\theta = 31°$; $B = 12.2$, $\theta = -35°$

(c) $C = \sqrt{C_x^2 + C_y^2 + C_z^2}$; the angle between C and $C = 5.39$; $\theta = 42°$, $\phi = 236°$

the z axis is $\theta = \cos^{-1}(C_z/C)$; the angle with the x axis is $\phi = \tan^{-1}(C_y/C_z)$

21* • If $A = 5\,i - 4\,j$ and $B = -7.5\,i + 6\,j$, write an equation relating A to B.

Note $B_x = -1.5A_x$ and $B_y = -1.5A_y$. Consequently, $B = -1.5A$.

25* • How is it possible for a particle moving at constant speed to be accelerating? Can a particle with constant velocity be accelerating at the same time?

A particle moving at constant speed in a circular path is accelerating (the direction of the velocity vector is changing). If a particle is moving at constant velocity, it is not accelerating.

29* •• As a bungee jumper approaches the lowest point in her drop, she loses speed as she continues to move downward. Draw the velocity vectors of the jumper at times t_1 and t_2, where $\Delta t = t_2 - t_1$ is small. From your drawing find the direction of the change in velocity $\Delta v = v_2 - v_1$, and thus the direction of the acceleration vector. The sketch is shown.

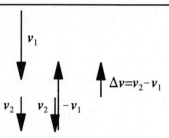

33* • A particle moving at 4.0 m/s in the positive x direction is given an acceleration of 3.0 m/s^2 in the positive y direction for 2.0 s. The final speed of the particle is ____. (a) -2.0 m/s (b) 7.2 m/s (c) 6.0 m/s (d) 10 m/s (e) None of the above

1. Find the final velocity vector $v = v_x i + a_y t j = 4.0 \text{ m/s } i + 6.0 \text{ m/s } j$
2. Find the magnitude of the velocity vector $v = 7.2$ m/s   (b)

37* •• A particle moves in the xy plane with constant acceleration. At time zero, the particle is at $x = 4$ m, $y = 3$ m, and has velocity $v = 2 \text{ m/s } i - 9 \text{ m/s } j$. The acceleration is given by the vector $a = 4 \text{ m/s}^2 i + 3 \text{ m/s}^2 j$. (a) Find the velocity vector at $t = 2$ s. (b) Find the position vector at $t = 4$ s. Give the magnitude and direction of the position vector.

(a) $v(t) = v_0 + at$ $v(2) = 10 \text{ m/s } i - 3 \text{ m/s } j$
(b) $r(t) = r_0 + v_0 t + \frac{1}{2}at^2$ $r(4) = 44 \text{ m } i - 9 \text{ m } j$; $r = 44.9$ m, $\theta = -11.6°$

41* • A river is 0.76 km wide. The banks are straight and parallel (Figure 3-34). The current is 5.0 km/h and is parallel to the banks. A boat has a maximum speed of 3 km/h in still water. The pilot of the boat wishes to go on a straight line from A to B, where AB is perpendicular to the banks. The pilot should (a) head directly across the river. (b) head 68° upstream from the line AB. (c) head 22° upstream from the line AB. (d) give up—the trip from A to B is not possible with this boat. (e) None of the above.

(d) give up; since the speed of the stream is greater than that of the boat in still water, it will always drift down stream.

45* •• Two boat landings are 2.0 km apart on the same bank of a stream that flows at 1.4 km/h. A motorboat makes the round trip between the two landings in 50 minutes. What is the speed of the boat relative to the water?

1. Find the speeds upstream and downstream $v_{up} = v - 1.4$ km/h; $v_{down} = v + 1.4$ km/h
2. Express total time in terms of v_{up} and v_{down} $T = (2 \text{ km})/v_{up} + (2 \text{ km})/v_{down} = 5/6$ h
3. Obtain equation for v and solve $v^2 - 4.8 v - 1.96 = 0$; $v = -0.378, 5.18$
4. Select physically acceptable solution $v = 5.18$ km/h

49* ••• Airports A and B are on the same meridian, with B 624 km south of A. Plane P departs airport A for B at the same time that an identical plane, Q, departs airport B for A. A steady 60 km/h wind is blowing from the south 30° east of north. Plane Q arrives at airport A 1 h before plane P arrives at airport B. Determine the airspeeds of the two planes (assuming that they are the same) and the heading of each plane.

Let v_{AP} and v_{AQ} be the velocity of P and Q relative to air; v_A is the velocity of the air (wind); v_P and v_Q the velocities of planes P and Q relative to the ground. Chose i and j as before.

1. Set east-west component of v_P and v_Q equal to zero.	$v_{AP} \sin \theta_P = (-60 \text{ km/h}) \sin 30° = -30 \text{ km/h} = v_{AQ} \sin \theta_Q;$
Let θ be the angle relative to north direction.	$\theta_P = \theta_Q = \theta$
2. Find north-south components of v_P and v_Q; these are the ground speeds of P and Q	$v_P = v_{AP} \cos \theta - (60 \text{ km/h}) \cos 30° = v_{AP} \cos \theta - 52 \text{ km/h}$ $v_Q = v_{AQ} \cos \theta + 52 \text{ km/h}$
3. Write the time difference in terms of $v_{AQ} = v_{AP} = v$	$1 \text{ h} = (624 \text{ km})[1/(v \cos \theta - 52 \text{ km/h}) + 1/(v \cos \theta + 52 \text{ km/h})]$
4. Solve for $v \cos \theta$	$v \cos \theta = 260 \text{ km/h}$
5. Note that $v \sin \theta = -30 \text{ km/h}$	$\theta = \tan^{-1}(30/260) = -6.58°$, i.e., 6.58° west of north
6. Solve for v	$v = v_{AP} = v_{AQ} = (260 \text{ km/h})/\cos 6.58° = 261.7 \text{ km/h}$

53* • A projectile was fired at 35° above the horizontal. At the highest point in its trajectory, its speed was 200 m/s. The initial velocity had a horizontal component of (a) 0. (b) (200 cos 35°) m/s. (c) (200 sin 35°) m/s. (d) (200/cos 35°) m/s. (e) 200 m/s.

(e) At the highest point, the speed is the horizontal component of the initial velocity.

57* • A pitcher throws a fast ball at 140 km/h toward home plate, which is 18.4 m away. Neglecting air resistance (not a good idea if you are the batter), find how far the ball drops because of gravity by the time it reaches home plate.

1. Find time of flight	$t = [18.4 \text{ m}/ (140/3.6) \text{ m/s}] = 0.473 \text{ s}$
2. Find distance dropped in 0.473 s	$d = \frac{1}{2}gt^2 = 1.1 \text{ m}$

61* •• If the tree in Example 3-11 is 50 m away and the monkey hangs from a branch 10 m above the muzzle position, what is the minimum initial speed of the dart if it is to hit the monkey before hitting the ground? We shall assume that the muzzle of the dart gun is 1.2 m above ground.

1. Find the time for monkey to fall to ground	$t = (2h/g)^{1/2} = (22.4/9.81)^{1/2} = 1.51 \text{ s}$
2. Projection angle $\theta = \tan^{-1}(10/50)$	$\theta = 11.3°$
3. $v_x = (50/1.51) \text{ m/s}; v_0 = v_x/\cos 11.3°$	$v_0 = [50/(1.51 \cos 11.3°)] \text{ m/s} = 33.8 \text{ m/s}$

65* •• At half its maximum height, the speed of a projectile is three-fourth its initial speed. What is the angle of the initial velocity vector with respect to the horizontal?

1. Find v_{0y}^2 in terms of H and g	$v_{0y}^2 = 2gH$
2. Find v_y^2 at $H/2$; use Equ. 2-15	$v_y^2 = 2gH - 2gH/2 = gH$
3. Find v^2 at $H/2$	$v^2 = v_y^2 + v_x^2 = gH + v_x^2 = (9/16)[2gH + v_x^2]$
4. Solve for v_x^2 in terms of v_{0y}^2	$7v_x^2 = v_{0y}^2; v_{0y} = \sqrt{7} \, v_x$
5. $\theta = \tan^{-1}(v_{0y}/v_x)$	$\theta = \tan^{-1}(\sqrt{7}) = 69.3°$

69* •• A stone thrown horizontally from the top of a 24-m tower hits the ground at a point 18 m from the base of the tower. (a) Find the speed at which the stone was thrown. (b) Find the speed of the stone just before it hits the ground.

(a) 1. Find the time to fall 24 m　　　　$t = \sqrt{2H/g} = 2.21$ s

2. Find $v_x = v_0$　　　　$v_x = v_0 = \Delta x/t = (18/2.21)$ m $= 8.14$ m/s

(b) 1. Find v_y at 2.21 s　　　　$v_y = gt = 21.7$ m/s

2. Find v_f　　　　$v_f = (v_x^2 + v_y^2)^{1/2} = 23.2$ m/s

73* •• Compute $dR/d\theta$ from Equation 3-22 and show that setting $dR/d\theta = 0$ gives $\theta = 45°$ for the maximum range.
$d[(v_0^2\sin 2\theta)/g]/d\theta = (v_0^2/g)d(\sin 2\theta)/d\theta = (2v_0^2/g)\cos 2\theta$; set equal to 0; $2\theta = 90°$, $\theta = 45°$.

77* ••• A girl throws a ball at a vertical wall 4 m away (Figure 3-38). The ball is 2 m above ground when it leaves the girl's hand with an initial velocity of $v_0 = (10\,i + 10\,j)$ m/s. When the ball hits the wall, the horizontal component of its velocity is reversed; the vertical component remains unchanged. Where does the ball hit the ground? Note: The wall acts like a mirror. We shall determine the range, neglecting the wall. Then consider the mirror-like reflection.

1. Use Equ. 2-14 to find t, time of flight　　　$0 = (2 + 10t - \frac{1}{2}gt^2)$ m; $t = 2.22$ s

2. Find the distance Δx　　　$\Delta x = 10t$ m $= 22.2$ m

3. Consider wall reflection; x from wall $= \Delta x - 4$ m　　　The ball will land 18.2 m from the wall

---

81* •• The coach throws a baseball to a player with an initial speed of 20 m/s at an angle of 45° with the horizontal. At the moment the ball is thrown, the player is 50 m from the coach. At what speed and in what direction must the player run to catch the ball at the same height at which it was released?

1. Use Equ. 3-22 to find the range of ball　　　$R = (400/9.81)$ m $= 40.8$ m

2. Find the distance the runner must cover　　　$\Delta x = 50 - 40.8 = 9.2$ m

3. Find the time of flight　　　$\Delta t = [40.8/(20\cos 45°)] = 2.88$ s

4. Find the speed of runner toward coach　　　$v = \Delta x/\Delta t = (9.2/2.88)$ m/s $= 3.19$ m/s

---

85* ••• If a bullet that leaves the muzzle of a gun at 250 m/s is to hit a target 100 m away at the level of the muzzle, the gun must be aimed at a point above the target. How far above the target is that point?

1. Find θ_0 using Equ. 3-22　　　$\sin 2\theta_0 = 981/250^2 = 0.0157$; $\theta_0 = 0.45°$ (or 89.55°)

2. $y = R\tan\theta_0$　　　$y = 0.785$ m (disregard $\theta_0 = 89.55°$ as unrealistic)

89* • True or false: (a) The magnitude of the sum of two vectors must be greater than the magnitude of either vector. (b) If the speed is constant, the acceleration must be zero. (c) If the acceleration is zero, the speed must be constant.

(a) False (b) False (c) True

93* •• The automobile path shown in Figure 3-41 is made up of straight lines and arcs of circles. The automobile starts from rest at point A. After it reaches point B, it travels at constant speed until it reaches point E. It comes to rest at point F. (a) At the middle of each segment (AB, BC, CD, DE, and EF), what is the direction of the velocity vector? (b) At which of these points does the automobile have an acceleration? In those cases, what is the direction of the acceleration? (c) How do the magnitudes of the acceleration compare for segments BC and DE?

Let y direction be north, x direction be east.

(*a*) AB, **v** directed north; BC, **v** directed northeast; CD, **v** directed east; DE, **v** directed southeast; EF, **v** directed south.

(*b*) AB, **a** to north; BC, **a** to southeast; CD, **a** = 0; DE, **a** to southwest; EF, **a** to north.

(*c*) The magnitudes are equal.

97* •   The position vector of a particle is given by $r = 5t\,i + 10t\,j$, where *t* is in seconds and *r* is in meters. (*a*) Draw the path of the particle in the *xy* plane. (*b*) Find **v** in component form and then find its magnitude.

(*b*) $v = dr/dt$ $v = (5\,i + 10\,j)$ m/s; $v = \sqrt{125}$ m/s = 11.2 m/s

(*a*) The path is shown in the figure

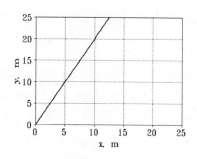

101* ••   Estimate how far you can throw a ball if you throw it (*a*) horizontally while standing on level ground; (*b*) at $\theta = 45°$ while standing on level ground; (*c*) horizontally from the top of a building 12 m high; (*d*) at $\theta = 45°$ from the top of a building 12 m high.

The answer will, of course, depend on the initial speed. Assume an initial speed of 90 km/h = 25 m/s.

(*a*) Assume the throwing arm is 1.5 m off the ground. Time to drop 1.5 m under gravity is $t = (3/9.81)^{½}$ s = 0.55 s. In that time the ball travels a distance $\Delta x = (25$ m/s$)(0.55$ s$)$ = 14 m.

(*b*) Neglect the initial height of the arm. Use Equ. 3-22 to get range $R = v^2/g$ = 64 m.

(*c*) Now the distance to drop is 13.5 m and $t = 1.66$ s. $\Delta x = 25 \times 1.66$ m = 41 m.

(*d*) Now we cannot neglect the difference between initial and final elevations. So we use Equ. 2-14 to determine *t*. We find $t = 4.25$ s and $\Delta x = (25/\sqrt{2})(4.25)$ m = 75 m.

105* ••   A particle moves in the *xy* plane with constant acceleration. At $t = 0$ the particle is at $r_1 = 4$ m $i + 3$ m j, with velocity v_1. At $t = 2$ s the particle has moved to $r_2 = 10$ m $i - 2$ m j, and its velocity has changed to $v_2 = (5\,i - 6\,j)$ m/s. (*a*) Find v_1. (*b*) What is the acceleration of the particle? (*c*) What is the velocity of the particle as a function of time? (*d*) What is the position vector of the particle as a function of time?

Note: For constant acceleration: $v_{av} = ½(v_1 + v_2)$, $a = (v_2 - v_1)/\Delta t$, $v_{av} = (r_2 - r_1)/\Delta t$; $v(t) = v_1 + at$; $r(t) = r_1 + v_1 t + ½at^2$.

(*a*) 1. Use the third of the above expressions

$v_{av} = [(10 - 4)$ m $i + (-2 - 3)$m $j]/2$ s
$= 3$ m/s $i - 2.5$ m/s j

2. Use the first of the above to find v_1

$v_1 = 2v_{av} - v_2$; $v_1 = (6 - 5)$ m/s $i + (-5 + 6)$ m/s j
$= (i + j)$ m/s

(*b*) Use the second of the above relations

$a = [(5\,i - 6\,j - i - j)/2]$ m/s$^2 = (2\,i - 3.5\,j)$ m/s^2

(*c*) Use the fourth relation above

$v(t) = [i + j + (2\,i - 3.5\,j)t]$ m/s

(*d*) Use the last of the above relations

$r(t) = [4\,i + 3\,j + (i + j)t + ½(2\,i - 3.5\,j)t^2]$ m

109*•• A baseball hit toward center field will land 72 m away unless caught first. At the moment the ball is hit, the center fielder is 98 m away. He uses 0.5 s to judge the flight of the ball, then races to catch it. The ball's speed as it leaves the bat is 35 m/s. Can the center fielder catch the ball before it hits the ground?

1. Find the angle θ when ball leaves bat from Equ. 3-22

$\sin 2\theta = (72 \times 9.81/35^2) = 0.5766;\ 2\theta = 35.2°$ or $125.2°$
(a) $\theta = 17.6°$ (line drive); (b) $\theta = 62.6°$ (high fly)

(a) 1. Find time of flight (line drive); use Equ. 2-12

$t = 2(35 \sin 17.6°)/9.81$ s $= 2.16$ s

  2. Time to reach ball = 2.16 s – 0.5 s = 1.66 s

$v_{min} = (26/1.66)$ m/s $= 15.7$ m/s ; faster than the world record. The ball cannot be caught.

(b) 1. Find time of flight (high fly); use Equ. 2-12

$t = 2(35 \sin 62.6°)/9.81$ s $= 6.34$ s

  2. Find v_{min} for fly ball

$v_{min} = 4.45$ m/s; the ball can be caught

113*••• Two balls are thrown with equal speeds from the top of a cliff of height H. One ball is thrown upward and an angle α above the horizontal. The other ball is thrown downward at an angle β below the horizontal. Show that each ball strikes the ground with the same speed, and find that speed in terms of H and the initial speed v_0.

1. Note that from Equ. 2-15, $v_y^2 = v_{0y}^2 + 2gH$, regardless of direction (up or down).
2. $v^2 = v_x^2 + v_y^2$, and since $v_x = v_{0x}$, $v^2 = v_{0x}^2 + v_{0y}^2 + 2gH = v_0^2 + 2gH$, for any angle, positive or negative.
3. $v = (v_0^2 + 2gH)^{1/2}$.

CHAPTER 4

Newton's Laws

Note: For all problems we shall take the upward direction as positive unless otherwise stated.

1* •• How can you tell if a particular reference frame is an inertial reference frame?

If Newton's first law is obeyed, the reference frame is an inertial reference frame.

5* • If an object is acted upon by a single known force, can you tell in which direction the object will move using no other information?

No; only its acceleration is known (assuming one knows the mass).

9* • A particle of mass m is traveling at an initial speed $v_0 = 25.0$ m/s. It is brought to rest in a distance of 62.5 m when a net force of 15.0 N acts on it. What is m? (*a*) 37.5 kg (*b*) 3.00 kg (*c*) 1.50 kg (*d*) 6.00 kg (*e*) 3.75 kg

$a = v^2/2s = F/m$; $m = v^2/2Fs$ $m = 3.00$ kg; (*b*)

13* • A force $F = 6$ N $i - 3$ N j acts on an object of mass 1.5 kg. Find the acceleration a. What is the magnitude a?

$a = F/m = (4i - 2j)$ m/s²; $a = (16 + 4)^{\frac{1}{2}}$ m/s² = 4.46 m/s².

17* • A 4-kg object is subjected to two forces, $F_1 = 2$ N $i - 3$ N j and $F_2 = 4$ n $i - 11$ N j. The object is at rest at the origin at time $t = 0$. (*a*) What is the object's acceleration? (*b*) What is its velocity at time $t = 3$ s? (*c*) Where is the object at time $t = 3$ s?

(*a*) Find the net force, $F = F_1 + F_2$ $F = (2i + 4i)$ N − $(3j + 11j)$ N = $(6i − 14j)$ N

     $a = F/m$ $a = (1.5i − 3.5j)$ m/s²

(*b*) $v = at$ $v = (4.5i − 10.5j)$ m/s

(*c*) $r = v_{av} t$; $v_{av} = \frac{1}{2}v(3$ s) $r = (6.75i − 15.75j)$ m

21* • On the moon, the acceleration due to gravity is only about 1/6 of that on earth. An astronaut whose weight on earth is 600 N travels to the lunar surface. His mass as measured on the moon will be (*a*) 600 kg. (*b*) 100 kg. (*c*) 61.2 kg. (*d*) 9.81 kg. (*e*) 360 kg.

(*c*) The mass is (600 N)/(9.81 m/s²) = 61.2 kg, and remains the same on the moon.

25* •• Caught without a map again, Hayley lands her spacecraft on an unknown planet. Visibility is poor, but she finds someone on a local communications channel and asks for directions to Earth. "You are already on Earth," is the reply, "Wait there and I'll be right over." Hayley is suspicious, however, so she drops a lead ball of mass

76.5 g from the top of her ship, 18 m above the surface of the planet. It takes 2.5 s to reach the ground. (*a*) If Hayley's mass is 68.5 kg, what is her weight on this planet? (*b*) Is she on Earth?

(*a*) Use $s = \frac{1}{2}at^2$ to find accel. of gravity, g' $g' = (2 \times 18/2.5^2)$ m/s^2 = 5.76 m/s^2

$\quad\quad w = mg'$ $w = (68.5 \times 5.76)$ N = 395 N

(*b*) Evidently, she is not on Earth.

29* •   A baseball player hits a ball with a bat. If the force with which the bat hits the ball is considered the action force, what is the reaction force? (*a*) The force the bat exerts on the batter's hands. (*b*) The force on the ball exerted by the glove of the person who catches it. (*c*) The force the ball exerts on the bat. (*d*) The force the pitcher exerts on the ball while throwing it. (*e*) Friction, as the ball rolls to a stop.

(*c*)

33* •   A vertical spring of force constant 600 N/m has one end attached to the ceiling and the other to a 12-kg block resting on a horizontal surface so that the spring exerts an upward force on the block. The spring is stretched by 10 cm. (*a*) What force does the spring exert on the block? (*b*) What is the force that the surface exerts on the block?

Draw a free body diagram and show the forces acting on the 12-kg block. F_k is the force exerted by the spring; $W = mg$ is the weight of the block; F_n is the normal force exerted by the horizontal surface.

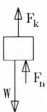

(*a*) $F_k = -kx$, where x is the extension of the spring $F_k = (600 \times 0.1)$ N = 60 N

(*b*) $\Sigma F = 0$; solve for F_n $F_n = (12 \times 9.81 - 60)$ N = 57.7 N

37* •   A clothesline is stretched taut between two poles. Then a wet towel is hung at the center of the line. Can the line remain horizontal? Explain.

No; the tension in the line must have a vertical component to support the weight of the towel.

41* •   A hovering helicopter of mass m_h is lowering a truck of mass m_t. If the truck's downward speed is increasing at the rate $0.1g$, what is the tension in the supporting cable? (*a*) $1.1m_t g$ (*b*) $m_t g$ (*c*) $0.9m_t g$ (*d*) $1.1 (m_h + m_t)g$

(*e*) $0.9 (m_h + m_t)g$

(*c*)

45* ••   A student has to escape from his girlfriend's dormitory through a window that is 15.0 m above the ground. He has a 24-m rope, but it will break when the tension exceeds 360 N, and the student weighs 600 N. The student will be injured if he hits the ground with a speed greater than 8 m/s. (*a*) Show that he cannot safely slide down the rope. (*b*) Find a strategy using the rope that will permit the student to reach the ground safely.

(*a*) $a = T/m - g$; $v = (2as)^{1/2}$ $a = -0.4g$; $v = -(2 \times 0.4 \times 9.81 \times 15)^{1/2}$ m/s = -10.8 m/s

(*b*) Double the rope and drop last 3 m   Now $v = -(2 \times 9.81 \times 3)^{1/2} = -7.7$ m/s

---

49* ••   A 1000-kg load is being moved by a crane. Find the tension in the cable that supports the load as (*a*) it is accelerated upward at 2 m/s^2, (*b*) it is lifted at constant speed, and (*c*) it moves upward with speed decreasing by 2 m/s each second.

(*a*), (*b*), and (*c*) $T = m(a - g)$; $g = -9.81$ m/s^2 (*a*) $T = 11810$ N; (*b*) $T = 9810$ N; (*c*) $T = 7810$ N

53* • A box slides down a frictionless inclined plane. Draw a diagram showing the forces acting on the box. For each force in your diagram, indicate the reaction force.

The forces acting on the box are its weight, **W**, and the normal reaction force of the inclined plane on the box, F_n. The reaction forces are indicated with primes.

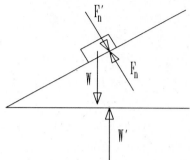

57* •• A horizontal force of 100 N pushes a 12-kg block up a frictionless incline that makes an angle of 25° with the horizontal. (*a*) What is the normal force that the incline exerts on the block? (*b*) What is the acceleration of the block?

Draw a free-body diagram for the box. Let *x* point to the right and up along the plane with *y* in the direction of F_n.

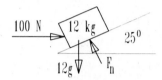

(*a*) Write $\Sigma F_y = 0$ and solve for F_n $F_n - mg \cos 25° - (100 \text{ N}) \sin 25° = 0; F_n = 149$ N

(*b*) Write $\Sigma F_x = ma$ and solve for *a* $(100 \text{ N}) \cos 25° - mg \sin 25° = ma; a = 3.41$ m/s²

61* • Suppose you are standing on a scale in a descending elevator as it comes to a stop on the ground floor. Will the scale's report of your weight be high, low, or correct?

It will be high because the acceleration is upward.

65* •• A 2-kg block hangs from a spring balance calibrated in newtons that is attached to the ceiling of an elevator (Figure 4-37). What does the balance read when (*a*) the elevator is moving up with a constant velocity of 30 m/s; (*b*) the elevator is moving down with a constant velocity of 30 m/s; (*c*) the elevator is ascending at 20 m/s and gaining speed at a rate of 10 m/s²? From $t = 0$ to $t = 2$ s, the elevator moves up at 10 m/s. Its velocity is then reduced uniformly to zero in the next 2 s, so that it is at rest at $t = 4$ s. Describe the reading of the balance during the interval $0 < t < 4$ s.

(*a*) $a = 0; F = mg = 19.6$ N. (*b*) $F = 19.6$ N. (*c*) $a = 10$ m/s²; $F = (2 \text{ kg})[(10 + 9.81) \text{ m/s}^2] = 39.6$ N.

For $0 \le t \le 2$ s, $F = 19.6$ N; for $2 \text{ s} \le t \le 4$ s, $a = -5$ m/s², so $F = (2 \text{ kg})(4.81 \text{ m/s}^2) = 9.62$ N.

69* •• Two blocks are in contact on a frictionless, horizontal surface. The blocks are accelerated by a horizontal force *F* applied to one of them (Figure 4-40). Find the acceleration and the contact force for (*a*) general values of $F, m_1,$ and m_2, and (*b*) for $F = 3.2$ N, $m_1 = 2$ kg, and $m_2 = 6$ kg.

(*a*) $a = F/(m_1 + m_2)$ and contact force $F_c = m_2 a$. So $F_c = Fm_2/(m_1 + m_2)$.

(*b*) Substitute numerical values in above expressions. $a = (3.2/8)$ m/s² = 0.4 m/s²; $F_c = (3.2 \times 6/8)$ N = 2.4 N.

73* •• Two climbers on an icy (frictionless) slope, tied together by a 30-m rope, are in the predicament shown in Figure 4-43. At time $t = 0$, the speed of each is zero, but the top climber, Paul (mass 52 kg), has taken one step too many and his friend Jay (mass 74 kg) has dropped his pick. (*a*) Find the tension in the rope as Paul falls and his speed just before he hits the ground. (*b*) If Paul unhooks his rope after hitting the ground, find Jay's speed as he hits the ground.

(*a*) In Problem 4-72 we derived

$a = g[(m_2 - m_1 \sin \theta)/(m_1 + m_2)]$,

$T = g[m_1 m_2/(m_1 + m_2)](1 + \sin \theta)$;

$v = (2as)^{1/2}$

$a = (9.81)[(52 - 74 \sin 40°)/126]$ m/s^2 = 0.345 m/s^2

$T = (9.81)(52 \times 74/126)(1 + \sin 40°) = 492$ N

$v = (2 \times 0.345 \times 20)^{1/2}$ m/s = 3.71 m/s

(*b*) 1. Find the distance Jay slides

     2. Find the acceleration of Jay

     3. Find $v_J = (2a_J s_J)^{1/2}$

$s_J = (25$ m$)/\sin 40° = 33.9$ m

$a_J = g \sin 40° = 6.31$ m/s^2

$v_J = (2 \times 33.9 \times 6.31)^{1/2}$ m/s = 20.7 m/s

77* ••   A heavy rope of length 5 m and mass 4 kg lies on a frictionless horizontal table. One end is attached to a 6-kg block. At the other end of the rope, a constant horizontal force of 100 N is applied. (*a*) What is the acceleration of the system? (*b*) Give the tension in the rope as a function of position along the rope.

(*a*) $F = (m_1 + m_2)a$

(*b*) *T* at 6-kg mass is 60 N. *x* is distance from 6-kg

$a = (100/10)$ m/s^2 = 10 m/s^2

$T(x) = [60 + (40/5)x]$ N = 60 + 8*x* N

81* ••   The apparatus in Figure 4-50 is called an *Atwood's machine* and is used to measure acceleration due to gravity *g* by measuring the acceleration of the two blocks. Assuming a massless, frictionless pulley amd a massless string, show that the magnitude of the acceleration of either body and the tension in the string are

$$a = \frac{m_1 - m_2}{m_1 + m_2} g \quad \text{and} \quad T = \frac{2m_1 m_2}{m_1 + m_2} g$$

1. Apply $\Sigma F = ma$ for each block

2. Add the two equations and solve for *a*

3. Use the expression for *a* to obtain the equation for *T*

$m_1 g - T = m_1 a;\ T - m_2 g = m_2 a$

85* ••• The acceleration of gravity *g* can be determined by measuring the time *t* it takes for a mass m_2 in an Atwood's machine to fall a distance *L*, starting from rest. (*a*) Find an expression for *g* in terms of m_1, m_2, *L*, and *t*. (*b*) Show that if there is a small error in the time measurement *dt*, it will lead to an error in the determination of *g* by an amount *dg* given by $dg/g = -2\ dt/t$. If $L = 3$ m and m_1 is 1 kg, find the value of m_2 such that *g* can be measured with an accuracy of ±5% with a time measurement that is accurate to 0.1 s. Assume that the only significant uncertainty in the measurement is the time of fall.

(*a*) Use $a = 2L/t^2$; from Problem 4-81, $g = a[(m_1 + m_2)/(m_1 - m_2)] = (2L/t^2)[(m_1 + m_2)/(m_1 - m_2)]$.

(*b*) Differentiate with respect to *t*. $dg/dt = -2g/t$ or $dg/g = -2dt/t$.

With $dg/g = \pm0.05$, $dt/t = \mp0.025$ and $t = (0.1$ s $)/0.025 = 4$ s. Now find $a = 2L/t^2 = 0.375$ m/s^2 and solve for m_2 with $m_1 = 1$ kg, using the expression for *a* of Problem 81. One obtains $m_2 = 0.926$ kg or 1.08 kg.

89* •   A force of 12 N is applied to an object of mass *m*. The object moves in a straight line, with its speed increasing by 8 m/s every 2 s. Find *m*.

1. Find *a*

2. $F = ma$; $m = F/a$

$a = (8/2)$ m/s^2 = 4 m/s^2

$m = (12/4)$ kg = 3 kg

93* •   If you weigh 125 lb on the earth, what would your weight be in pounds on the moon, where the free-fall acceleration due to gravity is 5.33 ft/s²?

$g_E = 32$ ft/s²; $w_M = w_E(g_M/g_E)$ $w_M = (125 \times 5.33/32)$ lb = 20.8 lb

97* ••   A box of mass m_1 is pulled along a smooth horizontal surface by a force F exerted at the end of a rope that has a much smaller mass m_2, as shown in Figure 4-52. (*a*) Find the acceleration of the rope and block, assuming them to be one object. (*b*) What is the net force acting on the rope? (*c*) Find the tension in the rope at the point where it is attached to the block. (*d*) The diagram, with the rope perfectly horizontal along its length, is not quite accurate. Correct the diagram, and state how this correction affects your solution.

(*a*) $a = F/(m_1 + m_2)$. (*b*) $F_{net} = m_2a = Fm_2/(m_1 + m_2)$. (*c*) $T = m_1a = Fm_1/(m_1 + m_2)$.

(*d*) The rope sags and F has a vertical component and its horizontal component is less than F. Consequently, a will be somewhat smaller.

101* ••   The masses attached to each side of an Atwood's machine consist of a stack of five washers each of mass m, as shown in Figure 4-54. The tension in the string is T_0. When one of the washers is removed from the left side, the remaining washers accelerate and the tension decreases by 0.3 N. (*a*) Find m. (*b*) Find the new tension and the acceleration of each mass when a second washer is removed from the left side.

(*a*) Use the result of Problem 4-81 $T_0 = 5mg$; $T_0 - T = 5mg - (2 \times 4m \times 5m)g/9m = 0.3$ N

Solve for m $m = (0.3$ N$) \times 9/5g = 0.055$ kg = 55 g

(*b*) Use the results of Problem 4-81 $T = (2 \times 3 \times 5 \times 9.81m/8)$ N = 2.02 N; $a = (2/8)g$

                                                                         = 2.45 m/s²

105* •••   The pulley in an Atwood's machine is given an upward acceleration a, as shown in Figure 4-56. Find the acceleration of each mass and the tension in the string that connects them.

A constant upward acceleration has the same effect as an increase in the acceleration of gravity from g to $g + a$. Thus, the tension in the string is given by the expression of Problem 4-81 with g replaced by $(g + a)$:

$T = 2m_1m_2(g + a)/(m_1 + m_2)$. To find the acceleration of the mass m_2 consider the forces acting on m_2; they are the tension T and the weight m_2g. Thus, $a_2 = (T - m_2g)/m_2$. Substituting the expression just derived for T and simplifying, one obtains $a_2 = [(m_1 - m_2)g + 2m_1a]/(m_1 + m_2)$. To check these results consider the some limiting cases:

1. $a = 0$; 2. $m_1 = m_2 = m$; 3. $a = -g$ (free fall of the system).

1. Setting $a = 0$, T and a_2 reduce to the expression given in Problem 4-81, as they should.

2. Setting $m_1 = m_2 = m$, the tension in the string is $T = m(g + a)$, as expected, and $a_2 = a$, as expected.

3. Setting $a = -g$, as in free fall, $T = 0$, as expected, and $a_2 = -g$, as expected.

The expression for a_1 is the same as for a_2 with all subscripts interchanged, i.e.,

$a_1 = [(m_2 - m_1)g + 2m_2a]/(m_1 + m_2)$.

CHAPTER 5

Applications of Newton's Laws

1* • Various objects lie on the floor of a truck moving along a horizontal road. If the truck accelerates, what force acts on the objects to cause them to accelerate?

Force of friction between the objects and the floor of the truck.

5* • A block of mass m is at rest on a plane inclined at angle of 30° with the horizontal, as in Figure 5-38. Which of the following statements about the force of static friction is true? $(a) f_s > mg$ $(b) f_s > mg \cos 30°$ (c) $f_s = mg \cos 30°$ $(d) f_s = mg \sin 30°$ (e) None of these statements is true.

(d) f_s must equal in magnitude the component of the weight along the plane.

9* • A block of mass m is pulled at constant velocity across a horizontal surface by a string as in Figure 5-39. The magnitude of the frictional force is $(a) \mu_k mg.$ $(b) T \cos \theta.$ $(c) \mu_k(T - mg).$ $(d) \mu_k T \sin \theta.$ $(e) \mu_k(mg + T \sin \theta).$

(b) The net force is zero.

13* • The force that accelerates a car along a flat road is the frictional force exerted by the road on the car's tires. (a) Explain why the acceleration can be greater when the wheels do not spin. (b) If a car is to accelerate from 0 to 90 km/h in 12 s at constant acceleration, what is the minimum coefficient of friction needed between the road and tires? Assume that half the weight of the car is supported by the drive wheels.

(a) $\mu_s > \mu_k$; therefore f is greater if the wheels do not spin.

(b) 1. Draw the free-body diagram; the normal force on each pair of wheels is ½mg.

    2. Apply $\Sigma F = ma$ $f_s = ma = \mu_s F_n; F_n = ½mg$

    3. Solve for a $a = ½\mu_s g = (25 \text{ m/s}^2)/(12 \text{ s}) = 2.08 \text{ m/s}^2$

    4. Find μ_s $\mu_s = (2 \times 2.08/9.81) = 0.425$

17* • A 50-kg box that is resting on a level floor must be moved. The coefficient of static friction between the box and the floor is 0.6. One way to move the box is to push down on it at an angle θ with the horizontal. Another method is to pull up on the box at an angle θ with the horizontal. (a) Explain why one method is better than the other. (b) Calculate the force necessary to move the box by each method if $\theta = 30°$ and compare the answers with the result when $\theta = 0°$.

The free-body diagram for both cases, $\theta > 0$ and $\theta < 0$, is shown.

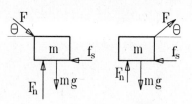

(a) $\theta > 0$ is preferable; it reduces F_n and therefore f_s.

(b) 1. Use $\Sigma F = ma$ to determine F_n $F \sin \theta + F_n - mg = 0$. $F_n = mg - F \sin \theta$

    2. $f_{s,max} = \mu_s F_n$ $f_{s,max} = \mu_s(mg - F \sin \theta)$

    3. To move the box, $F_x = F \cos \theta \geq f_{s,max}$ $F = \mu_s(mg - F \sin \theta)/\cos \theta;\ F = \dfrac{\mu_s mg}{\cos \theta + \mu_s \sin \theta}$

    4. Find F for $m = 50$ kg, $\mu_s = 0.6$, and $\theta = 30°$, $F(30°) = 252$ N, $F(-30°) = 520$ N, $F(0°) = 294$ N

      $\theta = -30°$, and $\theta = 0°$

21* ••   Returning to Figure 5-41, this time $m_1 = 4$ kg. The coefficient of static friction between the block and the incline is 0.4. (a) Find the range of possible values for m_2 for which the system will be in static equilibrium. (b) What is the frictional force on the 4-kg block if $m_2 = 1$ kg?

(a) 1. In Problem 5-20 we found

$$a = \frac{[m_2 - m_1(\sin \theta + \mu_k \cos \theta)]g}{m_1 + m_2};\ \text{set}\ a = 0. \qquad 0 = m_2 - m_1(\sin \theta \pm \mu_s \cos \theta)$$

    Note that f_s may point up or down the plane.

    2. Solve for m_2 with $m_1 = 4$ kg, $\mu_s = 0.4$ $m_2 = 3.39$ kg, 0.614 kg.

                                        $m_{2,max} = 3.39$ kg, $m_{2,min} = 0.614$ kg

(b) 1. Apply $\Sigma F = ma$; set $a = 0$ $m_2 g + f_s - m_1 g \sin \theta = 0$

    2. Solve for and find f_s $f_s = [(4.0 \times 0.5 - 1.0) \times 9.81]$ N $= 9.81$ N

25* ••   An automobile is going up a grade of 15° at a speed of 30 m/s. The coefficient of static friction between the tires and the road is 0.7. (a) What minimum distance does it take to stop the car? (b) What minimum distance would it take if the car were going down the grade?

The free-body diagram is shown for part (a).

For part (b), f_s points upward along the plane.

(a) 1. Apply $\Sigma F = ma$ $F_n = mg \cos \theta;\ ma_x = -mg \cos \theta - f_s$

    2. Replace $f_s = \mu_s F_n$ and solve for a_x $a_x = -(\sin \theta + \mu_s \cos \theta)g$

    3. Use $a_x = (v^2 - v_0^2)/2s$ and solve for s with

      $v^2 = 0$. $s = v_0^2/2g(\sin \theta + \mu_s \cos \theta);\ s = 49.1$ m

(b) Replace θ by $-\theta$ and solve for s. $s = v_0^2/2g(-\sin \theta + \mu_s \cos \theta);\ s = 110$ m

---

29* ••   Two blocks attached by a string slide down a 20° incline. The lower block has a mass of $m_1 = 0.25$ kg and a coefficient of kinetic friction $\mu_k = 0.2$. For the upper block, $m_2 = 0.8$ kg and $\mu_k = 0.3$. Find (a) the acceleration of the blocks and (b) the tension in the string.

1. Draw the free-body diagrams for each block. Since the coefficient of friction for the lower block is the smaller, the string will be under tension.

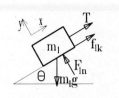 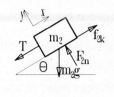

2. Apply $\Sigma F = ma$ to each block

$T + f_{1k} - m_1 g\sin\theta = m_1 a \qquad -T + f_{2k} - m_2 g\sin\theta = m_2 a$

$F_{1n} - m_1 g\cos\theta = 0 \qquad F_{2n} - m_2 g\cos\theta = 0$

3. Add the first pair of equations; use $f_k = \mu_k F_n$

$(m_1\mu_{1k} + m_2\mu_{2k})g\cos\theta - (m_1 + m_2)g\sin\theta = (m_1 + m_2)a$

4. Solve for a

$$a = \frac{(m_1\mu_{1k} + m_2\mu_{2k})\cos\theta - (m_1 + m_2)\sin\theta}{m_1 + m_2}\, g$$

5. Solve for T

$$T = \frac{m_1 m_2 (\mu_{2k} - \mu_{1k})g\cos\theta}{m_1 + m_2}$$

6. Substitute numerical values for the masses, friction coefficients, and θ to find a and T.

$a = -0.809$ m/s^2 (i.e., down the plane); $T = 0.176$ N

33* •• Answer the same questions as in Problem 32, only this time with a force F that pushes down on the block in Figure 5-44 at an angle θ with the horizontal.

(*a*) As in Problem 5-17, replace θ by $-\theta$ in the expression for F. One expects that F will increase with increasing magnitude of the angle since the normal component increases and tangential component decreases.

(*b*)

θ (degrees)	0	−10	−20	−30	−40	−50	−60
F (N)	240	272	327	424	631	1310	diverged

A plot of F versus the magnitude of θ is shown

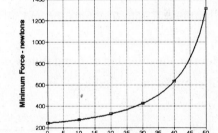

37* •• A 2-kg block sits on a 4-kg block that is on a frictionless table (Figure 5-47). The coefficients of friction between the blocks are $\mu_s = 0.3$ and $\mu_k = 0.2$. (*a*) What is the maximum force F that can be applied to the 4-kg block if the 2-kg block is not to slide? (*b*) If F is half this value, find the acceleration of each block and the force of friction acting on each block. (*c*) If F is twice the value found in (*a*), find the acceleration of each block.

(*a*) 1. Draw the free-body diagram

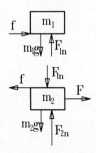

2. Apply $\Sigma F = ma$

$f_{s,max} = m_1 a_{max};\ F_{n1} = m_1 g;\ F_{max} - f_{s,max} = m_2 a_{max}$

3. Use $f_{s,max} = \mu_s F_{n1}$ and solve for a_{max} and F_{max}

$a_{max} = \mu_s g;\ F_{max} = (m_1 + m_2) g \mu_s;$

4. Evaluate a_{max} and F_{max}

$a_{max} = 2.94$ m/s^2, $F_{max} = 17.7$ N

(b) 1. The blocks move as a unit. The force on m_1 is

$a = F/(m_1 + m_2);\ a = 2.95$ m/s^2

$\quad m_1 a = f_s$.

$f_s = (2.95 \times 2)$ N $= 5.9$ N

(c) 1. If $F = 2F_{max}$ then m_1 slips on m_2.

$f = f_k = \mu_k m_1 g$

2. Apply $\Sigma F = ma$

$m_1 a_1 = f_k = \mu_k m_1 g;\ m_2 a_2 = F - \mu_k m_1 g$

3. Solve for and evaluate a_1 and a_2 for

$a_1 = \mu_k g;\ a_2 = (F - \mu_k g m_1)/m_2;$

$\quad F = 35.4$ N

$a_1 = 1.96$ m/s^2, $a_2 = 7.87$ m/s^2

41* ••• Lou has set up a kiddie ride at the Winter Ice Fair. He builds a right-angle triangular wedge, which he intends to push along the ice with a child sitting on the hypotenuse. If he pushes too hard, the kid will slide up and over the top, and Lou could be looking at a lawsuit. If he doesn't push hard enough, the kid will slide down the wedge, and the parents will want their money back. If the angle of inclination of the wedge is 40°, what are the minimum and maximum values for the acceleration that Lou must achieve? Use m for the child's mass, and μ_s for the coefficient of static friction between the child and the wedge.

1. Draw the free-body diagram. The diagram

for finding a_{min}; $f_s = f_{s,max} = \mu_s F_n$ and points

upward. To find a_{max}, reverse direction of f_s.

2. Apply $\Sigma F = ma$

$F_n \sin\theta - \mu_s F_n \cos\theta = ma;\ F_n \cos\theta + \mu_s F_n \sin\theta - mg = 0$

3. Use the second equation to solve for F_n

$F_n = mg/(\cos\theta + \mu_s \sin\theta)$

4. Substitute F_n into first equation and solve for

$\quad a = a_{min}$

$$a_{min} = g\frac{\sin\theta - \mu_s\cos\theta}{\cos\theta + \mu_s\sin\theta}$$

5. Reverse the direction of f_s and follow the

same procedure to find a_{max}.

$$a_{max} = g\frac{\sin\theta + \mu_s\cos\theta}{\cos\theta - \mu_s\sin\theta}$$

45* • A particle is traveling in a vertical circle at constant speed. One can conclude that the _____ is constant.

(a) velocity (b) acceleration (c) net force (d) apparent weight (e) none of the above

(e)

49* • A 0.75-kg stone attached to a string is whirled in a horizontal circle of radius 35 cm as in the conical pendulum of Example 5-10. The string makes an angle of 30° with the vertical. (a) Find the speed of the stone. (b) Find the tension in the string.

1. See Figure 5-28

2. Apply $\Sigma F = ma$

$T\cos\theta = mg;\ T\sin\theta = mv^2/r$

3. Solve for and evaluate v

$v = \sqrt{rg\tan\theta};\ v = 1.41$ m/s;

4. Evaluate T

$T = mg/\cos\theta;\ T = 8.5$ N

53* ••   Mass m_1 moves with speed v in a circular path of radius R on a frictionless horizontal table (Figure 5-52). It is attached to a string that passes through a frictionless hole in the center of the table. A second mass m_2 is attached to the other end of the string. Derive an expression for R in terms of m_1, m_2, and v.

1. Draw the free-body diagrams for the two masses

2. Apply $\Sigma F = ma$ $\qquad\qquad$ $T = m_1 v^2/R$

$\qquad\qquad\qquad\qquad\qquad\qquad\qquad\quad$ $T - m_2 g = 0$

3. Solve for R $\qquad\qquad\qquad\qquad$ $R = (m_1/m_2)v^2/g$

57* ••   A man swings his child in a circle of radius 0.75 m, as shown in the photo. If the mass of the child is 25 kg and the child makes one revolution in 1.5 s, what are the magnitude and direction of the force that must be exerted by the man on the child? (Assume the child to be a point particle.)

1. See Problem 5-49. In this problem T stands for the period.

2. $\tan \theta = v^2/rg = r\omega^2/g = 4\pi^2 r/gT^2$ $\qquad\qquad$ $\tan \theta = (4\pi^2 \times 0.75)/(9.81 \times 1.5^2) = 1.34$; $\theta = 53.3°$

(see Problem 5-49)

3. $F = mg/\cos \theta$ $\qquad\qquad\qquad\qquad$ $F = (25 \times 9.81/\cos 53.3°) = 410$ N

61* ••   A tether ball of mass 0.25 kg is attached to a vertical pole by a cord 1.2 m long. Assume the cord attaches to the center of the ball. If the cord makes an angle of 20° with the vertical, then (*a*) what is the tension in the cord? (*b*) What is the speed of the ball?

This problem is identical to Problem 5-49. With $r = L \sin \theta$, the relevant equations are: $T \cos \theta = mg$ and $v = \sqrt{Lg \sin \theta \tan \theta}$. Substituting the appropriate numerical values one obtains (*a*) $T = 2.61$ N, (*b*) $v = 1.21$ m/s.

65* •••   Revisiting the previous problem, (*a*) find the centripetal acceleration of the bead. (*b*) Find the tangential acceleration of the bead. (*c*) What is the magnitude of the resultant acceleration?

(*a*) Use the result of Problem 5-64 $\qquad\qquad$ $a_c = v^2/r = \dfrac{v_0^2}{r}\left(\dfrac{1}{1 + (\mu_k v_0/r)t}\right)^2$

(*b*) $a_t = -\mu_k v^2/r$ $\qquad\qquad\qquad\qquad$ $a_t = -\mu_k a_c$

(*c*) $a = (a_c^2 + a_t^2)^{\frac{1}{2}}$ $\qquad\qquad\qquad$ $a = a_c(1 + \mu_k^2)^{\frac{1}{2}}$, where a_c is given above.

69* •   Realizing that he has left the gas stove on, Aaron races for his car to drive home. He lives at the other end of a long, unbanked curve in the highway, and he knows that when he is traveling alone in his car at 40 km/h, he can just make it around the curve without skidding. He yells at his friends, "Get in the car! With greater mass, I can take the curve at higher speed!" Carl says, "No, that will make you skid at even lower speed." Bonita says, "The mass does not matter. Just get going!" Who is right?

Bonita is right.

73* ••   A curve of radius 150 m is banked at an angle of 10°. An 800-kg car negotiates the curve at 85 km/h without skidding. Find (*a*) the normal force on the tires exerted by the pavement, (*b*) the frictional force exerted by the pavement on the tires of the car, and (*c*) the minimum coefficient of static friction between the pavement and tires.

(*a*), (*b*) 1. Draw the free-body diagram

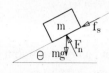

2. Apply $\Sigma F = ma$

$$F_n\sin\theta + f_s\cos\theta = mv^2/r \qquad (1)$$
$$F_n\cos\theta - f_s\sin\theta = mg \qquad (2)$$

3. Multiply (1) by sin θ, (2) by cos θ and add

$$F_n = (mv^2/r)\sin\theta + mg\cos\theta$$

4. Evaluate F_n and use (2) to evaluate f_s

$$F_n = 8245 \text{ N}; f_s = 1565 \text{ N}$$

(*c*) $\mu_{s,min} = f_s/F_n$

$$\mu_{s,min} = 0.19$$

77* •   How would you expect the value of *b* for air resistance to depend on the density of air?

The constant *b* should increase with density as more air molecules collide with the object as it falls.

81* •   A small pollution particle settles toward the earth in still air with a terminal speed of 0.3 mm/s. The particle

has a mass of 10^{-10} g and a retarding force of the form *bv*. What is the value of *b*?

When $v = v_t$, $bv = mg$, $b = mg/v_t$ $b = (10^{-13} \times 9.81/3 \times 10^{-4})$ kg/s $= 3.27 \times 10^{-9}$ kg/s

---

85* ••   An 800-kg car rolls down a very long 6° grade. The drag force for motion of the car has the form

$F_d = 100 \text{ N} + (1.2 \text{ N·s}^2/\text{m}^2)v^2$. What is the terminal velocity of the car rolling down this grade?

1. Draw the free-body diagram. Note that the car

moves at constant velocity, i.e., $a = 0$.

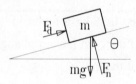

2. Apply $\Sigma F = ma$ $F_d = mg\sin\theta$

3. Use F_d as given, $m = 800$ kg, and $\theta = 6°$ $(100 + 1.2v^2)$ N $= 820$ N

4. Evaluate $v = v_t$ $v_t = 24.5$ m/s $= 88.2$ km/h

89* •••   Small spherical particles experience a viscous drag force given by Stokes' law: $F_d = 6\pi\eta rv$, where *r* is the

radius of the particle, *v* is its speed, and η is the viscosity of the fluid medium. (*a*) Estimate the terminal speed of

a spherical pollution particle of radius 10^{-5} m and density 2000 kg/m^3. (*b*) Assuming that the air is still and η is

1.8×10^{-5} N·s/m^2, estimate the time it takes for such a particle to fall from a height of 100 m.

Assume a spherical particle. Also, neglect the time required to attain terminal velocity; we will later confirm that

this assumption is justified.

(*a*) Using Stokes's law and $m = (4/3)\pi r^3\rho$ solve $v_t = (2r^2\rho g)/(9\eta) = 2.42$ cm/s

for v_t

(*b*) Find the time to fall 100 m at 2.42 cm/s $t = (10^4 \text{ cm})/(2.42 \text{ cm/s}) = 4.13 \times 10^3$ s $= 1.15$ h

In time $t' = 5v_t/g$, *v* is within 1% of v_t. [see $t' \approx 5v_t/g = 12$ ms $<<< 1.15$ h; neglect of t' is justified

Problem 5-88(*b*)]

93* •   On an icy winter day, the coefficient of friction between the tires of a car and a roadway might be reduced to

one-half its value on a dry day. As a result, the maximum speed at which a curve of radius *R* can be safely

negotiated is (*a*) the same as on a dry day. (*b*) reduced to 70% of its value on a dry day. (*c*) reduced to 50% of its value on a dry day. (*d*) reduced to 25% of its value on a dry day. (*e*) reduced by an unknown amount depending on the car's mass.

(*b*) $- v_{max} = (\mu_s g R)^{1/2}$; therefore $v'_{max} = v_{max}/\sqrt{2}$.

97* ••   An 800-N box rests on a plane inclined at 30° to the horizontal. A physics student finds that she can prevent the box from sliding if she pushes with a force of at least 200 N parallel to the surface. (*a*) What is the coefficient of static friction between the box and the surface? (*b*) What is the greatest force that can be applied to the box parallel to the incline before the box slides up the incline?

(*a*) 1. Draw the free-body diagram.

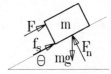

2. Apply $\Sigma F = ma$ $F + f_s - mg \sin \theta = 0$; $F_n = mg \cos \theta$

3. Use $f_{s,max} = \mu_s F_n$ and solve for and find μ_s $\mu_s = \tan \theta - F/(mg \cos \theta)$; $\mu_s = 0.289$

(*b*) 1. Find $f_{s,max}$ from part (*a*) $f_{s,max} = mg \sin \theta - F = 400 \text{ N} - 200 \text{ N} = 200 \text{ N}$

2. Reverse the direction of $f_{s,max}$ and evaluate F $F = mg \sin \theta + f_{s,max} = 400 \text{ N} + 200 \text{ N} = 600 \text{ N}$

101* ••   An object with a mass of 5.5 kg is allowed to slide from rest down an inclined plane. The plane makes an angle of 30° with the horizontal and is 72 m long. The coefficient of kinetic friction between the plane and the object is 0.35. The speed of the object at the bottom of the plane is (*a*) 5.3 m/s. (*b*) 15 m/s. (*c*) 24 m/s. (*d*) 17 m/s. (*e*) 11 m/s.

1. Draw the free-body diagram

2. Apply $\Sigma F = ma$ $mg \sin \theta - \mu_k F_n = ma$;

     $F_n - mg \cos \theta = 0$

3. Solve for a $a = g(\sin \theta - \mu_k \cos \theta)$

4. Use $v^2 = 2as$ $v = \sqrt{2(g \sin \theta - \mu_k g \cos \theta)s} = 16.7$ m/s

5. (*d*) is correct

105* ••   A flat-topped toy cart moves on frictionless wheels, pulled by a rope under tension T. The mass of the cart is m_1. A load of mass m_2 rests on top of the cart with a coefficient of static friction μ_s. The cart is pulled up a ramp that is inclined at an angle θ above the horizontal. The rope is parallel to the ramp. What is the maximum tension T that can be applied without making the load slip?

1. Draw the free-body diagrams for the two objects.

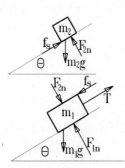

1. m_2 is accelerated by f_s. Apply $\Sigma F = ma$ $\qquad$ $F_{n2} = m_2 g \cos\theta$; $\mu_s m_2 g \cos\theta - m_2 g \sin\theta = m_2 a_{max}$

2. Solve for a_{max} $\qquad$ $a_{max} = g(\mu_s \cos\theta - \sin\theta)$

3. The masses move as single unit. Apply $\Sigma F = ma$ $\qquad$ $T - (m_1 + m_2)g\sin\theta = (m_1 + m_2)g(\mu_s \cos\theta - \sin\theta)$

4. Solve for T $\qquad$ $T = (m_1 + m_2)g\mu_s \cos\theta$

109* •• In an amusement-park ride, riders stand with their backs against the wall of a spinning vertical cylinder. The floor falls away and the riders are held up by friction. If the radius of the cylinder is 4 m, find the minimum number of revolutions per minute necessary to prevent the riders from dropping when the coefficient of static friction between a rider and the wall is 0.4.

1. Apply $\Sigma F = ma$ $\qquad$ $F_n = mr\omega^2$; $f_{s,max} = \mu_s F_n = mg$

2. Solve for and evaluate ω $\qquad$ $\omega = (g/\mu_s r)^{\frac{1}{2}}$; $\omega = 2.476$ rad/s $= 23.6$ rpm

---

CHAPTER **6**

Work and Energy

1* • True or false: (*a*) Only the net force acting on an object can do work. (*b*) No work is done on a particle that remains at rest. (*c*) A force that is always perpendicular to the velocity of a particle never does work on the particle.

(*a*) False (*b*) True (*c*) True

5* • An object moves in a circle at constant speed. Does the force that accounts for its acceleration do work on it? Explain.

No; $dW = \textbf{F} \cdot d\textbf{r}$ and here $\textbf{F}$ is perpendicular to $d\textbf{r}$.

9* • A 6-kg box is raised from rest a distance of 3 m by a vertical force of 80 N. Find (*a*) the work done by the force, (*b*) the work done by gravity, and (*c*) the final kinetic energy of the box.

(*a*) $W = Fy$	$W = 3 \times 80 \text{ J} = 240 \text{ J}$
(*b*) $W_g = -mgy$	$W_g = -(6)(9.81)(3) \text{ J} = -177 \text{ J}$
(*c*) $K = W + W_g$	$K = 63 \text{ J}$

13* •• A 3-kg particle is moving with a speed of 2 m/s when it is at $x = 0$. It is subjected to a single force F_x that varies with position as shown in Figure 6-30. (*a*) What is the kinetic energy of the particle when it is at $x = 0$? (*b*) How much work is done by the force as the particle moves from $x = 0$ to $x = 4$ m? (*c*) What is the speed of the particle when it is at $x = 4$ m?

(*a*) $K(0) = \frac{1}{2}mv^2$	$K(0) = \frac{1}{2}(3)(4) \text{ J} = 6 \text{ J}$
(*b*) $W = $ area under curve	$W = 2 \times 6 \text{ J} = 12 \text{ J}$
(*c*) $K(4) = K(0) + W = \frac{1}{2}mv^2$	$\frac{1}{2}(3)v^2 = 18 \text{ J}; \ v = 3.46 \text{ m/s}$

17* •• A 3-kg object is moving with a speed of 2.40 m/s in the *x* direction when it passes the origin. It is acted on by a single force F_x that varies with *x* as shown in Figure 6-32. (*a*) What is the work done by the force from $x = 0$ to $x = 2$ m? (*b*) What is the kinetic energy of the object at $x = 2$ m? (*c*) What is the speed of the object at $x = 2$ m? (*d*) What is the work done on the object from $x = 0$ to $x = 4$ m? (*e*) What is the speed of the object at $x = 4$ m?

(*a*) $W = $ area under curve; count squares, each square $= 0.125$ J	Between $x = 0$ and $x = 2$ there are about 22 squares, corresponding to $W = 2.75$ J

(b) $K = K_i + W$ $K = \tfrac{1}{2}(3)(2.4)^2$ J + 2.75 J = 11.4 J

(c) $v = (2K/m)^{1/2}$ $v = (2 \times 11.4/3)^{1/2} = 2.76$ m/s

(d) As in (a) count squares   Net number of squares = 28; W = 3.5 J

(e) Proceed as in (b) and (c) $K = (8.64 + 3.5)$ J = 12.14 J; $v = 2.84$ m/s

---

21* • An 85-kg cart is deposited on a 1.5-m platform after being rolled up an incline formed by a plank of length L that has been laid from the lower level to the top of the platform. (Assume the rolling is equivalent to sliding without friction.) (a) Find the force parallel to the incline needed to push the cart up without acceleration for L = 3, 4, and 5 m. (b) Calculate directly from Equation 6-15 the work needed to push the cart up the incline for each value of L. (c) Since the work found in (b) is the same for each value of L, what advantage, if any, is there in choosing one length over another?

(a) $F = mg \sin\theta = mg(1.5/L)$; $mg = 834$ N $L = 3$ m, $F = 417$ N; $L = 4$ m, $F = 313$ N; $L = 5$ m, $F = 250$ N

(b) $W = F \cdot s$ For $L = 3$ m, $W = 3 \times 417$ J = 1.25 kJ; same for other L

(c) Choosing longer length means one can exert a smaller force

---

25* • Find $A \cdot B$ for the following vectors: (a) $A = 3\,i - 6\,j$, $B = -4\,i + 2\,j$; (b) $A = 5\,i + 5\,j$, $B = 2\,i - 4\,j$; and (c) $A = 6\,i + 4\,j$, $B = 4\,i - 6\,j$.

(a), (b), (c) Use Equ. 6-12   (a) $A \cdot B = -24$; (b) $A \cdot B = -10$; (c) $A \cdot B = 0$.

29* •• When a particle moves in a circle with constant speed, the magnitudes of its position vector and velocity vector are constant. (a) Differentiate $r \cdot r = r^2 = $ constant with respect to time to show that $v \cdot r = 0$ and therefore $v \perp r$. (b) Differentiate $v \cdot v = v^2 = $ constant with respect to time and show that $a \cdot v = 0$ and therefore $a \perp v$. What do the results of (a) and (b) imply about the direction of a? (c) Differentiate $v \cdot r = 0$ with respect to time and show that $a \cdot r + v^2 = 0$ and therefore $a_r = -v^2/r$.

(a) $(d/dt)(r \cdot r) = r \cdot (dr/dt) + (dr/dt) \cdot r = 2v \cdot r = 0$. Therefore $v \perp r$.

(b) $(d/dt)(v \cdot v) = 2a \cdot v = 0$. Therefore $a \perp v$.

(c) The above implies that the component of a in the plane formed by r and v is colinear with r.

(d) $(d/dt)(v \cdot r) = v \cdot (dr/dt) + r \cdot (dv/dt) = v^2 + r \cdot a = 0$. Therefore, $a_r = -v^2/r$.

33* • The engine of a car operates at constant power. The ratio of acceleration of the car at a speed of 60 km/h to that at 30 km/h (neglecting air resistance) is (a) $\frac{1}{2}$. (b) $1/\sqrt{2}$. (c) $\sqrt{2}$. (d) 2.

(a) $mav = $ constant; $av = $ constant and $a \propto 1/v$.

37* • Fluffy has just caught a mouse, and decides that the only decent thing to do is to bring it to the bedroom so that his human roommate can admire it when she wakes up. A constant horizontal force of 3 N is enough to drag the mouse across the rug at constant speed v. If Fluffy's force does work at the rate of 6 W, (a) what is her speed, v? (b) How much work does Fluffy do in 4 s?

(a) Use Equ. 6-17 $v = (6/3)$ m/s = 2 m/s

(b) $W = Pt$ $W = (6 \times 4)$ J = 24 J

41* •• At a speed of 20 km/h, a 1200-kg car accelerates at 3 m/s^2 using 20 kW of power. How much power must be expended to accelerate the car at 2 m/s^2 at a speed of 40 km/h?

$P = Fv = mav$ $P = (1200 \times 2 \times 40/3.6)$ W = 26.7 kW

45* •• A 4.0-kg object initially at rest at $x = 0$ is accelerated at constant power of 8.0 W. At $t = 9.0$ s, it is at $x = 36.0$ m. Find its speed at $t = 6.0$ s and its position at that instant.

$v = (2Pt/m)^{\frac{1}{2}}$; $x(6) = x(9)(6/9)^{3/2}$ $v = (2 \times 8 \times 6/4)^{\frac{1}{2}} = 4.9$ m/s; $x(6) = (36)(6/9)^{3/2} = 19.6$ m

---

49* • A woman runs up a flight of stairs. The gain in her gravitational potential energy is U. If she runs up the same stairs with twice the speed, what will be her gain in potential energy? (*a*) U (*b*) $2U$ (*c*) $U/2$ (*d*) $4U$ (*e*) $U/4$

(*a*) The change in U does not depend on speed, only on difference in elevation.

53* • When you climb a mountain, is the work done on you by gravity different if you take a short, steep trail instead of a long, gentle trail? If not, why do you find one trail easier?

No; a steeper trail requires more effort per step, but fewer steps.

57* • Water flows over Victoria Falls, which is 128 m high, at an average rate of 1.4×10^6 kg/s. If half the potential energy of this water were converted to electric energy, how much power would be produced by these falls?

$P = -\frac{1}{2}(dU/dt) = -\frac{1}{2}gh(dm/dt)$ $P = \frac{1}{2}(9.81 \times 128 \times 1.4 \times 10^6)$ W = 879 MW

61* •• A simple Atwood's machine uses two masses, m_1 and m_2 (Figure 6-34). Starting from rest, the speed of the two masses is 4.0 m/s at the end of 3.0 s. At that instant, the kinetic energy of the system is 80 J and each mass has moved a distance of 6.0 m. Determine the values of m_1 and m_2.

1. $K = \frac{1}{2}(m_1 + m_2)v^2$; solve for $m_1 + m_2$ $m_1 + m_2 = (2 \times 80/4^2)$ kg = 10 kg

2. $a = \dfrac{m_1 - m_2}{m_1 + m_2}g = v(t)/t$; solve for $m_1 - m_2$ $a = 1.33$ m/s^2; $m_1 - m_2 = 1.33(10/9.81)$ kg = 1.36 kg

3. Solve for m_1 and m_2 $m_1 = 5.68$ kg, $m_2 = 4.32$ kg

65* •• A potential-energy function is given by $U = C/x$, where C is a positive constant. (*a*) Find the force F_x as a function of x. (*b*) Is this force directed toward the origin or away from it? (*c*) Does the potential energy increase or decrease as x increases? (*d*) Answer parts (*b*) and (*c*) where C is a negative constant.

(*a*), (*b*), (*c*) $F_x = -dU/dx$ (*a*) $F_x = C/x^2$; (*b*) directed away from the origin
                                                          ($-x$ direction); (*c*) $U(x)$ increases with increasing x

(*d*) Repeat above with $C < 0$ Directed toward the origin; U increases as x increases

---

69* •• During a Dead Wait concert, Sharika and Chico, each of mass M, are attached to the ends of a light rope that is hung over two frictionless pulleys, as shown in Figure 6-38. A large gong of mass m is attached to the middle of the rope, between the pulleys, and Sharika and Chico beat it madly in lieu of the usual guitar solo. (*a*) Find the potential energy of the system as a function of the distance y to the center of the gong. (*b*) Find the value of y for which the potential energy function of the system is a minimum. (*c*) Find the equilibrium distance y_0 using the potential energy function. (*d*) Check your answer by applying Newton's laws to the gong.

(a) Set $U = 0$ for $y = 0$; note that each M is raised by $h = (d^2 + y^2)^{1/2} - d$ as m drops a distance y

$U(y) = 2Mg[(d^2 + y^2)^{1/2} - d] - mgy$

(b) Find dU/dy and set equal to zero

$dU/dy = 2Mgy/(d^2 + y^2)^{1/2} - mg$; $dU/dy = 0$ for $y = d(m/2M)/\sqrt{1 - (m/2M)^2}$

(c) Set net force on $m = 0$
    Solve for y

$F_{net} = mg - 2Mg \sin \theta$; $\sin \theta = y/(y^2 + d^2)^{1/2}$
$y = d(m/2M)/\sqrt{1 - (m/2M)^2}$

73* ••• A force is given by $F_x = Ax^{-3}$, where $A = 8$ N·m³. (a) For positive values of x, does the potential energy associated with this force increase or decrease with increasing x? (You can determine the answer to this question by imagining what happens to a particle that is placed at rest at some point x and is then released.) (b) Find the potential-energy function U associated with this force such that U approaches zero as x approaches infinity. (c) Sketch U versus x.

(a) Apply Equ. 6-21a

$U(x) = \frac{1}{2}A/x^2 + U_0$; for $x > 0$, U decreases as x increases

(b) Let $x \to \infty$ and set $U(x) \to 0$

$0 = 0 + U_0$; $U_0 = 0$; $U(x) = \frac{1}{2}(8$ N·m³$)/x^2$ J $= 4/x^2$ J

(c) The plot of $U(x)$ is shown

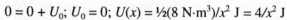
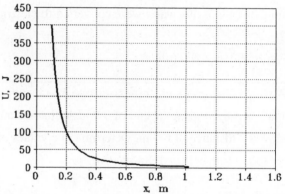

77* • Figure 6-39 shows two pulleys arranged to help lift a heavy load. A rope runs around two massless, frictionless pulleys and the weight w hangs from one pulley. You exert a force of magnitude F on the free end of the cord. (a) If the weight is to move up a distance h, through what distance must the force move? (b) How much work is done by the ropes on the weight? (c) How much work do you do? (This is an example of a simple machine in which a small force F_1 moves through a large distance x_1 in order to exert a large force $F_2 (= w)$ through a smaller distance $x_2 = h$.)

(a) If w moves a distance h, F moves a distance $2h$. (b) $W = wh$. (c) $W = F \times 2h$; note that the tension, F, in the string is $\frac{1}{2}w$. Thus $F \times 2h = wh$.

81* • A 2.4-kg object attached to a horizontal string moves with constant speed in a circle of radius R on a frictionless horizontal surface. The kinetic energy of the object is 90 J and the tension in the string is 360 N. Find R.

$\frac{1}{2}mv^2 = K$; $mv^2/R = T$; thus, $R = 2K/T$

$R = (2 \times 90/360)$ m $= 0.5$ m

85* •• Water from behind a dam flows through a large turbine at a rate of 1.5×10^6 kg/min. The turbine is located 50 m below the surface of the reservoir, and the water leaves the turbine with a speed of 5 m/s. (a) Neglecting

any energy dissipation, what is the power output of the turbine? (*b*) How many U.S. citizens would be supplied
with energy by this dam if each citizen uses 3×10^{11} J of energy per year?

(*a*) $E_{init}/kg = gh = E_{fin}/kg = \frac{1}{2}v_f^2 + W_{el}/kg$;

$P_{el} = MW_{el}/\Delta t$ $P_{el} = [1.5 \times 10^6(9.81 \times 50 - \frac{1}{2} \times 5^2)/(60)]$ W = 12 MW

(*b*) 3×10^{11} J/y = 9.51 kW $N = 12 \times 10^6/9.51 \times 10^3 = 1261$

89* ••   A 3-kg particle starts from rest at $x = 0$ and moves under the influence of a single force $F_x = 6 + 4x - 3x^2$,
where F_x is in newtons and x is in meters. (*a*) Find the work done by the force as the particle moves from $x = 0$ to
$x = 3$ m. (*b*) Find the power delivered to the particle when it is at $x = 3$ m.

(*a*) $W = \int F_x dx$ $W = \int_0^3 F(x)\,dx = (6x + 2x^2 - x^3)\Big|_0^3 = 9$ J

(*b*) Find v from $v = (2K/m)^{\frac{1}{2}}$, $K = W$ $v = (18/3)^{\frac{1}{2}}$ m/s = 2.45 m/s

Use $P = Fv$ $P = (-9\ \text{N})(2.45\ \text{m/s}) = -22$ W

93* ••   A rope of length L and mass per unit length of μ lies coiled on the floor. (*a*) What force F is required to hold
one end of the rope a distance $y < L$ above the floor as shown in Figure 6-42? (*b*) Find the work required to lift
one end of the rope from the floor to a height $l < L$ by integrating $F\,dy$ from $y = 0$ to $y = l$.

(*a*) The mass supported is μy; $F = \mu yg$. (*b*) $W = \int_0^l \mu gy\,dy = \mu g \dfrac{l^2}{2}$.

CHAPTER 7

Conservation of Energy

1* •• What are the advantages and disadvantages of using the conservation of mechanical energy rather than Newton's laws to solve problems?

It is generally simpler, involving only scalars, but some details cannot be obtained, e.g., trajectories.

5* • A woman on a bicycle traveling at 10 m/s on a horizontal road stops pedaling as she starts up a hill inclined at 3.0° to the horizontal. Ignoring friction forces, how far up the hill will she travel before stopping? (*a*) 5.1 m (*b*) 30 m (*c*) 97 m (*d*) 10.2 m (*e*) The answer depends on the mass of the woman.

(*c*) $h = v^2/2g = 50/9.81$ m $= 5.1$ m; $d = (5.1/\sin 3.0°)$ m $= 97.4$ m.

9* • The 3-kg object in Figure 7-18 is released from rest at a height of 5 m on a curved frictionless ramp. At the foot of the ramp is a spring of force constant $k = 400$ N/m. The object slides down the ramp and into the spring, compressing it a distance x before coming momentarily to rest. (*a*) Find x. (*b*) What happens to the object after it comes to rest?

(*a*) $U_i = mgh$; $U_f = \tfrac{1}{2}kx^2$; $K_i = K_f = 0$; use Equ. 7-6 $x = \sqrt{2mgh/k} = 0.858$ m

(*b*) The spring will accelerate the mass and it will then retrace its path, rising to a height of 5 m.

13* • A stone is projected horizontally with a speed of 20 m/s from a bridge 16 m above the surface of the water. What is the speed of the stone as it strikes the water?

Take $U = 0$ at surface of water.

$E_i = \tfrac{1}{2}mv_i^2 + mgh = E_f = \tfrac{1}{2}mv_f^2$ $v_f = (v_i^2 + 2gh)^{\frac{1}{2}} = 26.7$ m/s

---

17* •• The system shown in Figure 7-19 is initially at rest when the lower string is cut. Find the speed of the objects when they are at the same height.

1. $E_i = E_f$; $\Delta U + \Delta K = 0$ $\Delta U = (2 \times 0.5 - 3 \times 0.5) \times 9.81$ J $= -4.91$ J

2. $K = \tfrac{1}{2}mv^2 = -\Delta U$, $m = 5$ kg; find v $v = (2 \times 4.91/5)^{\frac{1}{2}}$ m/s $= 1.40$ m/s

---

21* •• A 2.4-kg block is dropped from a height of 5.0 m onto a spring of spring constant 3955 N/m. When the block is momentarily at rest, the spring has compressed by 25 cm. Find the speed of the block when the compression of the spring is 15.0 cm.

1. Let ΔU_g be the change in gravitational potential energy U; use Equ. 7-6 $\Delta U_g = -mg(h + x) = \frac{1}{2}mv^2 + \frac{1}{2}kx^2$

2. Find v at $x = 0.15$ m $v = \sqrt{2g(h + x) - kx^2/2} = \pm 8.0$ m/s

---

25* ••   A stone is thrown upward at an angle of 53° above the horizontal. Its maximum height during the trajectory is 24 m. What was the stone's initial speed?

1. $\frac{1}{2}mv_y^2 = mgh$; solve for and evaluate v_y $v_y = (2gh)^{\frac{1}{2}} = 21.7$ m/s

2. $v = v_y/\sin 53°$ $v = 27.2$ m/s

29* ••   A pendulum consists of a 2-kg bob attached to a light string of length 3 m. The bob is struck horizontally so that it has an initial horizontal velocity of 4.5 m/s. For the point at which the string makes an angle of 30° with the vertical, what is (a) the speed? (b) the potential energy? (c) the tension in the string? (d) What is the angle of the string with the vertical when the bob reaches its greatest height?

(a) 1. Find h at $\theta = 30°$ $h = L(1 - \cos\theta)$; $h = 0.402$ m

  2. $v^2 = v_0^2 - 2gh$ $v = (4.5^2 - 2 \times 0.402 \times 9.81)^{\frac{1}{2}}$ m/s $= 3.52$ m/s

(b) $U = mgh$ $U = (2 \times 9.81 \times 0.402)$ J $= 7.89$ J

(c) $T = mg\cos\theta + mv^2/L$ $T = (2 \times 9.81 \times 0.866 + 2 \times 3.52^2/3)$ N $= 25.3$ N

(d) $\theta_0 = \cos^{-1}(1 - v^2/2gL)$ $\theta_0 = \cos^{-1}(1 - 4.5^2/2 \times 9.81 \times 3) = 49°$

33* ••   Walking by a pond, you find a rope attached to a tree limb 5.2 m off the ground. You decide to use the rope to swing out over the pond. The rope is a bit frayed but supports your weight. You estimate that the rope might break if the tension is 80 N greater than your weight. You grab the rope at a point 4.6 m from the limb and move back to swing out over the pond. (a) What is the maximum safe initial angle between the rope and the vertical so that it will not break during the swing? (b) If you begin at this maximum angle, and the surface of the pond is 1.2 m below the level of the ground, with what speed will you enter the water if you let go of the rope when the rope is vertical?

Here we must supply "your weight." We shall take your weight to be 650 N (about 145 lb).

(a) 1. $T_m = mg + mv^2/L = mg + 80$ N; $L = 4.6$ m $mv^2 = 4.6 \times 80$ J

  2. $\frac{1}{2}mv^2 = mgL(1 - \cos\theta)$ $\cos\theta = 1 - mv^2/2mgL = 0.938$; $\theta = 20.2°$

(b) $\Delta U = -[mgL(1 - \cos\theta) + mg\Delta h] = \frac{1}{2}mv^2$

  $v^2 = 2g[L(1 - \cos\theta) + 1.8$ m$]$; solve for v $v = 6.4$ m/s

---

37* •   A man stands on roller skates next to a rigid wall. To get started, he pushes off against the wall. Discuss the energy changes pertinent to this situation.

The kinetic energy of the man increases at the expense of metabolic (chemical) energy.

41* •   A 70-kg skater pushes off the wall of a skating rink, acquiring a speed of 4 m/s. (a) How much work is done on the skater? (b) What is the change in mechanical energy of the skater? (c) Discuss the conservation of energy as applied to the skater.

(a) No work is done on the skater. The displacement of the force exerted by the wall is zero.

(b) $\Delta K = \frac{1}{2}mv^2 = 560$ J.

(c) The increase in kinetic energy is at the loss of metabolic (chemical) energy.

45* • Discuss the energy considerations when you pull a box along a rough road.

Metabolic (chemical) energy is converted to thermal energy released through friction.

49* • The 2-kg block in Figure 7-28 slides down a frictionless curved ramp, starting from rest at a height of 3 m. The block then slides 9 m on a rough horizontal surface before coming to rest. (*a*) What is the speed of the block at the bottom of the ramp? (*b*) What is the energy dissipated by friction? (*c*) What is the coefficient of friction between the block and the horizontal surface?

(*a*) Use $\Delta K + \Delta U = 0$; $\Delta U = -mg\Delta h$; $\Delta h = 3$ m $\qquad K = mg\Delta h = \frac{1}{2}mv^2$; $v = \sqrt{2g\Delta h} = 7.67$ m/s

(*b*) At end, $K = 0$; so $W_f = K = mg\Delta h$ $\qquad\qquad W_f = 58.9$ J

(*c*) $W_f = f_k s = \mu_k mgs$; solve for and find μ_k $\qquad \mu_k = W_f/mgs = mg\Delta h/mgs = \Delta h/s = 1/3$

53* •• A particle of mass m moves in a horizontal circle of radius r on a rough table. It is attached to a horizontal string fixed at the center of the circle. The speed of the particle is initially v_0. After completing one full trip around the circle, the speed of the particle is $\frac{1}{2}v_0$. (*a*) Find the energy dissipated by friction during that one revolution in terms of m, v_0, and r. (*b*) What is the coefficient of kinetic friction? (*c*) How many more revolutions will the particle make before coming to rest?

(*a*) Here, $W_f = K_i - K_f$ since U is constant. Thus, $W_f = \frac{1}{2}m(v_0^2 - v_0^2/4) = (3/8)mv_0^2$.

(*b*) Distance traveled in one revolution is $2\pi r$; so $W_f = \mu_k mg(2\pi r)$ and $\mu_k = (3v_0^2)/(16\pi gr)$.

(*c*) Since in one revolution it lost $(3/4)K_i$, it will only require another 1/3 revolution to lose the remaining $K_i/4$.

57* ••• A block of mass m rests on a rough plane inclined at θ with the horizontal (Figure 7-30). The block is attached to a spring of constant k near the top of the plane. The coefficients of static and kinetic friction between the block and plane are μ_s and μ_k, respectively. The spring is slowly pulled upward along the plane until the block starts to move. (*a*) Obtain an expression for the extension d of the spring the instant the block moves. (*b*) Determine the value of μ_k such that the block comes to rest just as the spring is in its unstressed condition, i.e., neither extended nor compressed.

(*a*) The force exerted by the spring must equal the sum $mg \sin \theta + f_s$; hence, $d = (mg/d)(\sin \theta + \mu_s\cos \theta)$.

(*b*) Take $U_{grav} = 0$ at initial position of m. To meet the condition, $U_{grav} - W_f = \frac{1}{2}kd^2$, where $U_{grav} = mgd \sin \theta$, and $W_f = \mu_k mgd \cos \theta$. Solving for μ_k one finds $\mu_k = \tan \theta - \frac{1}{2}(1 + \mu_s \cot \theta)$.

61* • For the fusion reaction in Example 7-14, calculate the number of reactions per second that are necessary to generate 1 kW of power.

From Example 7-14, energy per reaction is 17.59 MeV $= 28.1 \times 10^{-13}$ J; to generate 1000 J/s then requires $(1000/28.1 \times 10^{-13})$ reactions $= 3.56 \times 10^{14}$ reactions.

65* •• A large nuclear power plant produces 3000 MW of power by nuclear fission, which converts matter into energy. (*a*) How many kilograms of matter does the plant consume in one year? (*b*) In a coal-burning power plant, each kilogram of coal releases 31 MJ of energy when burned. How many kilograms of coal are needed each year for a 3000-MW plant?

(*a*) 1. Find the energy produced per year $\qquad E = (3 \times 10^9 \times 3.16 \times 10^7)$ J $= 9.48 \times 10^{16}$ J

2. Find $m = E/c^2$ $\qquad\qquad m = (9.48 \times 10^{16}/9 \times 10^{16})$ kg $= 1.05$ kg

(*b*) Find m_{coal} $\qquad\qquad m_{coal} = (9.48 \times 10^{16}/3.1 \times 10^7)$ kg $= 3.06 \times 10^9$ kg

69* •   Our bodies convert internal chemical energy into work and heat at the rate of about 100 W, which is called our metabolic rate. (*a*) How much internal chemical energy do we use in 24 h? (*b*) The energy comes from the food that we eat and is usually measured in kilocalories, where 1 kcal = 4.184 kJ. How many kilocalories of food energy must we ingest per day if our metabolic rate is 100 W?

(*a*) $E = Pt$; $E = (100 \times 24 \times 60 \times 60)$ J = 8.64 MJ.

(*b*) $E = (8.64/4.184)$ Mcal = 2.065 Mcal = 2065 kcal.

73* ••   A T-bar tow is required to pull 80 skiers up a 600-m slope inclined at 15° above horizontal at a speed of 2.5 m/s. The coefficient of kinetic friction is 0.06. Find the motor power required if the mass of the average skier is 75 kg.

1. Find the force required	$F = m_{tot}g \sin\theta + \mu_k g m_{tot}\cos\theta$
2. Use $P = Fv$; evaluate P	$P = (600 \times 75 \times 9.81 \times 2.5)(\sin 15° + 0.06\cos 15°)$W
	$= 350$ kW

77* ••   The spring constant of a toy dart gun is 5000 N/m. To cock the gun the spring is compressed 3 cm. The 7-g dart, fired straight upward, reaches a maximum height of 24 m. Determine the energy dissipated by air friction during the dart's ascent. Estimate the speed of the projectile when it returns to its starting point.

1. Use conservation of energy	$\frac{1}{2}kx^2 = mgh + W_f$; $W_f = \frac{1}{2}kx^2 - mgh$
2. Evaluate W_f	$W_f = 0.602$ J
3. We assume that W_f same on descent (not true)	$\frac{1}{2}mv^2 = mgh - W_f$; $v = \sqrt{2gh - W_f/m} = 17.3$ m/s

81* ••   A car of mass 1500 kg traveling at 24 m/s is at the foot of a hill that rises 120 m in 2.0 km. At the top of the hill, the speed of the car is 10 m/s. Find the average power delivered by the car's engine, neglecting any frictional losses.

1. Find increase in mechanical energy; $\Delta E = \Delta K + \Delta U$	$\Delta E = [750(24^2 - 10^2) + 1500 \times 9.81 \times 120]$ J = 2.12 MJ
2. Assume constant a; then $v_{av} = 17$ m/s	$\Delta t = (2000/17)$ s = 118 s; $P_{av} = \Delta E/\Delta t = 18$ kW

85* ••   A roller-coaster car having a total mass (including passengers) of 500 kg travels freely along the winding frictionless track in Figure 7-34. Points A, E, and G are horizontal straight sections, all at the same height of 10 m above ground. Point C is at a height of 10 m above ground on a section sloped at an angle of 30°. Point B is at the top of a hill, while point D is at ground level at the bottom of a valley. The radius of curvature at each of these points is 20 m. Point F is at the middle of a banked horizontal curve of radius of curvature of 30 m, and at the same height of 10 m above the ground as points A, E, and G. At point A the speed of the car is 12 m/s. (*a*) If the car is just barely able to make it over the hill at point B, what is the height of that point above ground? (*b*) If the car is just barely able to make it over the hill at point B, what is the magnitude of the total force exerted on the car by the track at that point? (*c*) What is the acceleration of the car at point C? (*d*) What are the magnitude and direction of the total force exerted on the car by the track at point D? (*e*) What are the magnitude and direction of the total force exerted on the car by the track at point F? (*f*) At point G, a constant braking force is applied to the car, bring the car to a halt in a distance of 25 m. What is the braking force?
Take $U = 0$ at A, E, and G.

(a) Use Equ. 7-6; $U_i = 0$, $K_f = 0$ 　　　　　　　$mg\Delta h = \frac{1}{2}mv_A^2$; $\Delta h = 7.34$ m; $h = 17.34$ m

(b) $F = F_n = mg$ 　　　　　　　　　　　　$F = (500 \times 9.81)$ N $= 4905$ N

(c) $a = g \sin \theta$ 　　　　　　　　　　　$a = (9.81 \sin 30°)$ m/s^2 $= 4.905$ m/s^2

(d) 1. At D, $F = mg + mv^2/R$ directed up; find v 　$v_A = 0$, $v_D^2 = 2(17.34$ m$)g$ m^2/s^2; $mv_D^2/R = 867g$ N

　　2. Find F 　　　　　　　　　　　　$F = (867 + 500)g$ N $= 13.41$ kN

(e) F has two components, vertical and horizontal 　$F_v = mg = 4905$ N; $F_h = mv^2/R = 2400$ N

　　Find F and θ (angle with vertical) 　　　$\theta = \tan^{-1}(2.4/4.905) = 26°$;

　　　　　　　　　　　　　　　　　$F = (2400$ N$)/\sin 26° = 5461$ N

(f) Use $\Delta K + W_f$; $\Delta K = -\frac{1}{2}mv_A^2$; $W_f = F_{brake}d$ 　$F_{brake} = \frac{1}{2}mv_A^2/d = 1440$ N

89* •• A 2-kg block is released 4 m from a massless spring with a force constant $k = 100$ N/m that is fixed along a frictionless plane inclined at 30°, as shown in Figure 7-35. (a) Find the maximum compression of the spring. (b) If the plane is rough rather than frictionless, and the coefficient of kinetic friction between the plane and the block is 0.2, find the maximum compression. (c) For the rough plane, how far up the incline will the block travel after leaving the spring?

(a) 1. Use $\Delta E = 0$; let x (along plane) $= 0$ at 　　$mgL \sin \theta + mgx \sin \theta = \frac{1}{2}kx^2$; for $\theta = 30°$,

　　　equilibrium position of spring 　　　　　$k = 100$ N/m, $m = 2$ kg, $L = 4$ m,

　　　　　　　　　　　　　　　　　$50x^2 - 9.81x - 39.24 = 0$

　　2. Solve quadratic equation for x 　　　　$x = 0.989$ m

(b) 1. Express energy loss due to friction, W_f 　　$W_f = \mu_k mg \cos \theta(L + x)$

　　2. Apply work-energy theorem 　　　　　$mg \sin \theta(L + x) - \mu_k mg \cos \theta(L + x) = \frac{1}{2}kx^2$

　　3. Obtain quadratic equation for x and solve 　$50x^2 - 6.41x - 25.65 = 0$; $x = 0.783$ m

(c) 1. Now $E_{in} = \frac{1}{2}kx^2 = E_{mech} + W_f$ 　　　$\frac{1}{2}kx^2 - mg \sin \theta(x + L') - \mu_k mg \cos \theta(x + L') = 0$

　　2. Solve for L' with $x = 0.783$ m 　　　　$L' = 1.54$ m

93* •• On July 31, 1994, Sergey Bubka pole-vaulted over a height of 6.14 m. If his body was momentarily at rest at the top of the leap, and all the energy required to raise his body derived from his kinetic energy just prior to planting his pole, how fast was he moving just before takeoff? Neglect the mass of the pole. If he could maintain that speed for a 100-m sprint, how fast would he cover that distance? Since the world record for the 100-m dash is just over 9.8 s, what do you conclude about world-class pole-vaulters?

1. $\Delta E = 0$; $\frac{1}{2}mv^2 = mgh$ 　　　　　　$v = (2gh)^{\frac{1}{2}}$; $v = 10.96$ m/s

2. Find t for 100-m sprint at 10.96 m/s 　　　$t = 100/10.96$ s $= 9.12$ s

　A pole vaulter uses additional metabolic energy to

　raise himself on pole.

97* ••• The bob of a pendulum of length L is pulled aside so the string makes an angle θ_0 with the vertical, and the bob is then released. In Example 7-2, the conservation of energy was used to obtain the speed of the bob at the bottom of its swing. In this problem, you are to obtain the same result using Newton's second law. (a) Show that the tangential component of Newton's second law gives $dv/dt = -g \sin \theta$, where v is the speed and θ is the angle made by the string and the vertical. (b) Show that v can be written $v = L \, d\theta/dt$. (c) Use this result and the chain rule for derivatives to obtain $dv/dt = (dv/d\theta)(d\theta/dt) = (dv/d\theta)(v/L)$. (d) Combine the results of (a) and (c) to obtain $v \, dv = -gL \sin \theta \, d\theta$. (*e*) Integrate the left side of the equation in part (*d*) from $v = 0$ to the final speed v and the right side from $\theta = \theta_0$ to $\theta = 0$, and show that the result is equivalent to $v = \sqrt{2gh}$, where *h* is the original height of the bob above the bottom.

(*a*) $F_{tan} = -mg \sin \theta$; therefore $a = dv/dt = -g \sin \theta$

(*b*) For circular motion, $v = r\omega = L \, d\theta/dt$; $d\theta/dt = v/L$

(*c*) $dv/dt = (dv/d\theta)(d\theta/dt) = (v/L)(dv/d\theta)$

(*d*) $dv/d\theta = (L/v)(dv/dt) = -(L/v) g \sin \theta$; $v \, dv = -gL \sin \theta \, d\theta$

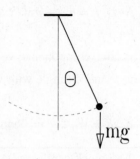

(*e*) $\int_0^v v \, dv = \int_{\theta_0}^0 -gL \sin \theta \, d\theta$; $\tfrac{1}{2}v^2 = gL(1 - \cos \theta_0)$

Note that $L(1 - \cos \theta_0) = h$; consequently, $v = \sqrt{2gh}$.

CHAPTER **8**

Systems of Particles and Conservation of Momentum

1* • Give an example of a three-dimensional object that has no mass at its center of mass.

A hollow sphere.

5* • Find the center of mass x_{cm} of the three masses in Figure 8-46.

Use Equ. 8-4 $\qquad\qquad x_{cm} = [(1 \times 1 + 2 \times 2 + 8 \times 4)/11]$ m $= 3.36$ m

---

9* •• The uniform sheet of plywood in Figure 8-49 has a mass of 20 kg. Find its center of mass. We shall consider this as two sheets, a square sheet of 3 m side length and mass m_1 and a rectangular sheet 1m × 2m with a mass of $-m_2$. Let coordinate origin be at lower left hand corner of the sheet. Let σ be the surface density of the sheet.

1. Find $x_{cm}(m_1)$, $y_{cm}(m_1)$ and $x_{cm}(m_2)$, $y_{cm}(m_2)$

By symmetry, $x_{cm}(m_1) = 1.5$ m, $y_{cm}(m_1) = 1.5$ m
and $x_{cm}(m_2) = 1.5$ m, $y_{cm}(m_2) = 2.0$ m

2. Determine m_1 and m_2

$m_1 = 9\sigma$ kg, $m_2 = 2\sigma$ kg

3. Use Equ. 8-4

$x_{cm} = (9\sigma \times 1.5 - 2\sigma \times 1.5)/7\sigma = 1.5$ m
$y_{cm} = (9\sigma \times 1.5 - 2\sigma \times 2.0)/7\sigma = 1.36$ m

---

13* ••• Find the center of mass of a thin hemispherical shell.

The element of area on the shell is $dA = 2\pi R^2 \sin\theta\, d\theta$, where R is the radius of the hemisphere. Let σ be the surface mass density. Then $M = \frac{1}{2}(4\pi R^2 \sigma) = 2\pi R^2 \sigma$. Use the same coordinates as in Problem 8-12, and apply Equ. 8-5:

$$z_{cm} = \frac{\int z\,\sigma dA}{M} = \frac{2\pi R^3 \sigma}{2\pi R^2 \sigma}\int_0^{\pi/2} \sin\theta\,\cos\theta\, d\theta = R/2.$$

17* • The two pucks in Problem 16 are lying on a frictionless table and connected by a spring of force constant k. A horizontal force F_1 is again exerted only on m_1 along the spring away from m_2. What is the magnitude of the acceleration of the center of mass? (a) F_1/m_1 (b) $F_1/(m_1 + m_2)$ (c) $(F_1 + k\Delta x)/(m_1 + m_2)$, where Δx is the amount the spring is stretched. (d) $(m_1 + m_2)F_1/m_1 m_2$

(b) by application of Equ. 8-10; the spring force is an internal force.

21* ••   A block of mass m is attached to a string and suspended inside a hollow box of mass M. The box rests on a scale that measures the system's weight. (a) If the string breaks, does the reading on the scale change? Explain your reasoning. (b) Assume that the string breaks and the mass m falls with constant acceleration g. Find the acceleration of the center of mass, giving both direction and magnitude. (c) Using the result from (b), determine the reading on the scale while m is in free fall.

(a) Yes; initially the scale reads $(M + m)g$; while m is in free fall, the reading is Mg.

(b) $a_{cm} = mg/(M + m)$, directed downward.

(c) $F_{net} = (M + m)g - (M + m)a_{cm} = Mg$.

25* •   True or false: (a) The momentum of a heavy object is greater than that of a light object moving at the same speed. (b) The momentum of a system may be conserved even when mechanical energy is not. (c) The velocity of the center of mass of a system equals the total momentum of the system divided by its total mass.

(a) True (for magnitude)   (b) True (inelastic collision)   (c) True

29* ••   Much early research in rocket motion was done by Robert Goddard, physics professor at Clark College in Worcester, Mass. A quotation from a 1921 editorial in the *New York Times* illustrates the public acceptance of his work: "That Professor Goddard with his 'chair' at Clark College and the countenance of the Smithonian Institution does not know the relation between action and reaction, and the need to have something better than a vacuum against which to react—to say that would be absurd. Of course, he only seems to lack the knowledge ladled out daily in high schools." The belief that a rocket needs something to push against was a prevalent misconception before rockets in space were commonplace. Explain why that belief is wrong.

Conservation of momentum does not require the presence of a medium such as air.

33* •   Figure 8-52 shows the behavior of a projectile just after it has broken up into three pieces. What was the speed of the projectile the instant before it broke up? (a) v_3 (b) $v_3/3$ (c) $v_3/4$ (d) $4v_3$ (e) $(v_1 + v_2 + v_3)/4$

(c) Use $p_i = p_f = mv_3 = 4mv_i$.

37* ••   A small object of mass m slides down a wedge of mass $2m$ and exits smoothly onto a frictionless table. The wedge is initially at rest on the table. If the object is initially at rest at a height h above the table, find the velocity of the wedge when the object leaves it.

$p_x = 0 = mv_x - 2mV$; $V = \frac{1}{2}v_x$; $v_x = \sqrt{2gh}$; $V = \sqrt{gh/2}$, directed opposite to that of the mass m.

41* ••   Repeat Problem 40 with the second, 3-kg block replaced by a block having a mass of 5 kg and moving to the right at 3 m/s.

(a) $K = K_1 + K_2$ $K = \frac{1}{2}[3 \times 25 + 5 \times 3^2]$ J $= 60$ J

(b) Use Equ. 8-13 $v_{cm} = (3 \times 5 + 5 \times 3)/8\, i$ m/s $= 3.75\, i$ m/s

(c) $v_{rel} = v - v_{cm}$ $v_{1,rel} = 1.25\, i$ m/s; $v_{2,rel} = -0.75\, i$ m/s

(d) $K_{rel} = K_{1,rel} + K_{2,rel}$ $K_{rel} = \frac{1}{2}(3 \times 1.25^2 + 5 \times 0.75^2)$ J $= 3.75$ J

(e) Find K_{cm} $K_{cm} = \frac{1}{2}(8 \times 3.75^2)$ J $= 56.25$ J $= K - K_{rel}$

45* •   A soccer ball of mass 0.43 kg leaves the foot of the kicker with an initial speed of 25 m/s. (a) What is the impulse imparted to the ball by the kicker? (b) If the foot of the kicker is in contact with the ball for 0.008 s, what is the average force exerted by the foot on the ball?

(a) Use Equ. 8-19 $I = mv = 10.75$ N·s

(b) Use Equ. 8-20 $F_{av} = 10.75/0.008$ N $= 1344$ N

49* ••   A 300-g handball moving with a speed of 5.0 m/s strikes the wall at an angle of 40° and then bounces off with the same speed at the same angle. It is in contact with the wall for 2 ms. What is the average force exerted by the ball on the wall?

1. Find Δv; $v_{xi} = v_0 \cos 40°$, $v_{xf} = -v_0 \cos 40°$ $\Delta v = 2 \times 5.0 \times \cos 40°$ m/s $= 7.66$ m/s

2. $F_{av} = m\Delta v/\Delta t$ $F_{av} = 0.3 \times 7.66/2 \times 10^{-3}$ N $= 1.15$ kN

53* •••   The great limestone caverns were formed by dripping water. (a) If water droplets of 0.03 mL fall from a height of 5 m at a rate of 10 per minute, what is the average force exerted on the limestone floor by the droplets of water? (b) Compare this force to the weight of a water droplet.

(a) 1. Find the mass of the droplet $m = (3 \times 10^{-5}$ L$)(1.0$ kg/L$) = 3 \times 10^{-5}$ kg

     2. Find v at impact $v = (2 \times 9.81 \times 5)^{\frac{1}{2}}$ m/s $= 9.9$ m/s

     3. Find $F_{av} = Nm\Delta v/\Delta t$; $N/\Delta t = (10/60)$ s^{-1} $F_{av} = (3 \times 10^{-5} \times 9.9/6)$ N $= 4.95 \times 10^{-5}$ N

(b) $w/F_{av} = 3 \times 9.81/4.95 = 6$ w is about six times the average force due to 10 drops

---

57* ••   Consider a perfectly inelastic collision of two objects of equal mass. (a) Is the loss of kinetic energy greater if the two objects have oppositely directed velocities of equal magnitude $v/2$, or if one of the two objects is initially at rest and the other has an initial velocity of v? (b) In which situation is the percentage loss in kinetic energy the greatest?

(a) Case 1: $K_i = 2(\frac{1}{2}mv^2/4)$, $K_f = 0$; $\Delta K = mv^2/4$. Case 2: $K_i = \frac{1}{2}mv^2$, $K_f = \frac{1}{2}[2m \times (v/2)^2] = mv^2/4$; $\Delta K = mv^2/4$. The energy losses are the same.

(b) The percentage loss is greatest (infinite) in case 1.

61* •   An 85-kg running back moving at 7 m/s makes a perfectly inelastic collision with a 105-kg linebacker who is initially at rest. What is the speed of the players just after their collision?

Use $p_i = p_f$; $v_f = v_i m_1/(m_1 + m_2)$ $v_f = 85 \times 7/190 = 3.13$ m/s

---

65* ••   Repeat Problem 64 with a second (illegal) muffin having a mass of 0.5 kg and moving to the right at 3 m/s.

(a) Use $p_i = p_f$; $v_f = (v_{1i}m_1+v_{2i}m_2)/(m_1 + m_2)$ $v_f = [(0.3 \times 5 + 0.5 \times 3)/0.8]$ m/s $= 3.75$ m/s

(b) 1. Transform to CM system; $u = v - v_{cm}$ $v_{cm} = 3.75$ m/s; $u_{1i} = 1.25$ m/s, $u_{2i} = -0.75$ m/s

     2. Use $u_f = -u_i$; transform back to lab system $u_{1f} = -1.25$ m/s, $u_{2f} = 0.75$ m/s; $v_{1f} = 2.5$ m/s,

                                                  $v_{2f} = 4.5$ m/s

---

69* ••   A block of mass $m_1 = 2$ kg slides along a frictionless table with a speed of 10 m/s. Directly in front of it, and moving in the same direction with a speed of 3 m/s, is a block of mass $m_2 = 5$ kg. A massless spring with spring constant $k = 1120$ N/m is attached to the second block as in Figure 8-54. (a) Before m_1 runs into the spring, what is the velocity of the center of mass of the system? (b) After the collision, the spring is compressed by a maximum amount Δx. What is the value of Δx? (c) The blocks will eventually separate again. What are the final velocities of the two blocks measured in the reference frame of the table?

(a) Use Equ. 8-13 $v_{cm} = (2 \times 10 + 5 \times 3)/7$ m/s $= 5$ m/s

(b) 1. At max. compression, $u_1 = u_2 = 0$; $K = K_{cm}$ $K_{cm} = \frac{1}{2} \times 7 \times 5^2$ J $= 87.5$ J

   2. Use conservation of energy: $\frac{1}{2}k\Delta x^2 = K_i - K_{cm}$ $K_i = (100 + 22.5)$ J $= 122.5$ J; $\frac{1}{2}k\Delta x^2 = 35$ J

   3. Solve for and evaluate Δx $\Delta x = (2 \times 35/1120)^{\frac{1}{2}}$ m $= 0.25$ m $= 25$ cm

(c) 1. Collision is elastic; find u_{1i} and u_{2i}; u_{1f} and u_{2f} $u_{1i} = 5$ m/s; $u_{2i} = -2$ m/s; $u_{1f} = -5$ m/s, $u_{2f} = 2$ m/s

   2. Transform to reference frame of table $v_{1f} = 0$ m/s. $v_{2f} = 7$ m/s

---

73* ••   A bullet of mass m_1 is fired with a speed v into the bob of a ballistic pendulum of mass m_2. The bob is attached to a very light rod of length L that is pivoted at the other end. The bullet is stopped in the bob. Find the minimum v such that the bob will swing through a complete circle.

1. Find v_i of bob + bullet to make complete circle $\frac{1}{2}(m_1+m_2)v_i^2 = 2gL(m_1+m_2)$; $v_i = 2\sqrt{gL}$

2. Use $p_i = p_f$ to find v $v = 2[(m_1 + m_2)/m_1]\sqrt{gL}$

77* ••   The light isotope of lithium, ^5Li, is unstable and breaks up spontaneously into a proton (hydrogen nucleus) and an α particle (helium nucleus). In this process, a total energy of 3.15×10^{-13} J is released, appearing as the kinetic energy of the two reaction products. Determine the velocities of the proton and α particle that arise from the decay of a ^5Li nucleus at rest. (*Note*: The masses of the proton and alpha particle are $m_p = 1.67 \times 10^{-27}$ kg and $m_\alpha = 4m_p = 6.68 \times 10^{-27}$ kg.)

1. Use $p_i = p_f = 0$ $4m_p v_\alpha = m_p v_p$; $v_\alpha = v_p/4$

2. Use energy conservation $\frac{1}{2}m_p v_p^2 + \frac{1}{2}m_\alpha v_\alpha^2 = (5/8)m_p v_p^2 = 3.15 \times 10^{-13}$ J

3. Solve for v_p and v_α $v_p = 1.74 \times 10^7$ m/s; $v_\alpha = 4.34 \times 10^6$ m/s

---

81* •   The coefficient of restitution for steel on steel is measured by dropping a steel ball onto a steel plate that is rigidly attached to the earth. If the ball is dropped from a height of 3 m and rebounds to a height of 2.5 m, what is the coefficient of restitution?

Find the ratio v_{rec}/v_{app} and use Equ. 8-31 $$e = \frac{v_{rec}}{v_{app}} = \sqrt{\frac{2gh_{rec}}{2gh_{app}}} = \sqrt{\frac{2.5}{3.0}} = 0.913$$

85* ••   A 2-kg block moving to the right with speed 5 m/s collides with a 3-kg block that is moving in the same direction at 2 m/s, as in Figure 8-55. After the collision, the 3-kg block moves at 4.2 m/s. Find (a) the velocity of the 2-kg block after the collision, and (b) the coefficient of restitution for the collision.

(a) Use $p_i = p_f$ $(2 \times 5 + 3 \times 2)$ kg·m/s $= (2v_{1f} + 3 \times 4.2)$ kg·m/s;
   $v_{1f} = 1.7$ m/s

(b) Use Equ. 8-31 $e = (4.2 - 1.7)/(5 - 2) = 0.833$

89* ••   Figure 8-57 shows the result of a collision between two objects of unequal mass. (a) Find the speed v_2 of the larger mass after the collision and the angle θ_2. (b) Show that the collision is elastic.

(a) 1. Use $p_i = p_f$; $3mv_0 = \sqrt{5}mv_0\cos\theta_1 + 2mv_2\cos\theta_2$;
   Note: $\sqrt{5}\sin\theta_1 = 2$, $\sqrt{5}\cos\theta_1 = 1$ $\sqrt{5}mv_0\sin\theta_1 = 2mv_2\sin\theta_2$

   2. Solve for θ_2 $2\cot\theta_2 = 3 - 2$; $\cot\theta_2 = 1$, $\theta_2 = 45°$

   3. Solve for v_2 $v_2 = \sqrt{2}v_0$

(b) Find K_i and K_f $K_i = 4.5mv_0^2$; $K_f = 2.5mv_0^2 + 2mv_0^2 = 4.5mv_0^2$; Q.E.D.

93* •• A particle with momentum p_1 in one dimension makes an elastic collision with a second particle of momentum $p_2 = -p_1$ in the center-of-mass reference frame. After the collision its momentum is p_1'. Write the total initial and final energies in terms of p_1 and p_1', and show that $p_1' = \pm p_1$. If $p_1' = -p_1$, the particle is merely turned around by the collision and leaves with the speed it had initially. What is the significance of the plus sign in your solution?

$K_{rel} = p_1^2/2m_1 + p_1^2/2m_2 = p_1^2(m_1 + m_2)/2m_1m_2$; $K_{cm} = (2p_1)^2/2(m_1 + m_2) = 2p_1^2/(m_1 + m_2)$; $K = K_{rel} + K_{cm}$, i.e., $K = (p_1^2/2)[(m_1^2 + 6m_1m_2 + m_2^2)/(m_1^2m_2 + m_1m_2^2)]$. In an elastic collision, $K_i = K_f$. Consequently, $(p_1')^2 = (p_1)^2$ and $p_1' = \pm p_1$. If $p_1' = p_1$, the particles do not collide.

97* •• The payload of a rocket is 5% of its total mass, the rest being fuel. If the rocket starts from rest and moves with no external forces acting on it, what is its final velocity if the exhaust velocity of its gas is 5 km/s?

1. No external forces are acting	Equ. 8-42 reduces to $v_f = -u_{ex} \ln(m_f/m_0)$
2. Evaluate v_f for $m_f/m_0 = 1/20$	$v_f = [5 \ln(20)]$ km/s $= 15$ km/s

101* • The condition necessary for the conservation of momentum of a given system is that (a) energy is conserved. (b) one object is at rest. (c) no external force acts. (d) internal forces equal external forces. (e) the net external force is zero.

(e)

105* • A 4-kg fish is swimming at 1.5 m/s to the right. He swallows a 1.2-kg fish swimming toward him at 3 m/s. Neglecting water resistance, what is the velocity of the larger fish immediately after his lunch?

Use Equ. 8-13 $v = v_{cm} = (4 \times 1.5 - 1.2 \times 3)/5.2$ m/s $= 0.462$ m/s

109* • Repeat Problem 106 for a 3-kg block moving at 6 m/s to the right and a 6-kg block moving at 3 m/s to the left.

(a) $K_t = K_1 + K_2$	$K_t = \frac{1}{2}(3 \times 6^2 + 6 \times 3^2)$ J $= 81$ J
(b) Use Equ. 8-13	$v_{cm} = (3 \times 6 - 6 \times 3)/9$ m/s $= 0$ m/s
(c) $K_{cm} = \frac{1}{2}Mv_{cm}^2$	$K_{cm} = 0$ J
(d) $K_{rel} = K_t - K_{cm}$	$K_{rel} = 81$ J

113* •• In World War I, the most awesome weapons of war were huge cannons mounted on railcars. Figure 8-60 shows such a cannon, mounted so that it will project a shall at an angle of 30°. With the car initially at rest, the cannon fires a 200-kg projectile at 125 m/s. Now consider a system composed of a cannon, shell, and railcar, all rolling on the track without frictional losses. (a) Will the total vector momentum of that system be the same (i.e., "conserved") before and after the shell is fired? Explain your answer in a few words. (b) If the mass of the railcar plus cannon is 5000 kg, what will be the recoil velocity of the car along the track after the firing? (c) The shell is observed to rise to a maximum height of 180 m as it moves through its trajectory. At this point, its speed is 80 m/s. On the basis of this information, calculate the amount of thermal energy produced by air friction on the shell on its way from firing to this maximum height.

(a) Momentum of system is not conserved; there is an external force, the vertical reaction force of rails.

(b) Use $p_{xi} = p_{xf}$ $200 \times 125 \cos 30° = 5000v_{rec}$; $v_{rec} = 4.33$ m/s

(c) 1. Find h without air friction; $h = v_y^2/2g$ $h = 199$ m

    2. $W_f = mg\Delta h + \frac{1}{2}m(v_{x0}^2 - v_x^2)$ $W_f = 200 \times 9.81 \times 19 + 100[(125 \cos 30°)^2 - 80^2)]$ J

                                                    $= 569$ kJ

117* •• Two particles of mass m and $4m$ are moving in a vacuum at right angles as in Figure 8-62. A force F acts on both particles for a time T. As a result, the velocity of the particle m is $4v$ in its original direction. Find the new velocity v' of the particle of mass $4m$.

1. Determine $FT = \Delta p$ $FT = 3mv\, \textbf{\textit{i}}$

2. Find $p_{4m}' = p_{4m}(0) + \Delta p$, and v' $p_{4m}' = 3mv\, \textbf{\textit{i}} - 4mv\, \textbf{\textit{j}}$; $v' = 0.75v\, \textbf{\textit{i}} - v\, \textbf{\textit{j}}$

121* •• A small car of mass 800 kg is parked behind a small truck of mass 1600 kg on a level road (Figure 8-64). The brakes of both the car and the truck are off so that they are free to roll with negligible friction. A man sitting on the tailgate of the truck shoves the car away by exerting a constant force on the car with his feet. The car accelerates at 1.2 m/s². (a) What is the acceleration of the truck? (b) What is the magnitude of the force exerted on either the truck or the car?

(a) $F_{ext} = 0$, $\therefore a_{cm} = 0$ 800×1.2 kg·m/s² $= 1600a_t$; $a_t = 0.6$ m/s^s

(b) $F = ma$ $F = 960$ N

125* •• Initially, mass $m = 1.0$ kg and mass M are both at rest on a frictionless inclined plane (Figure 8-66). Mass M rests against a spring that has a spring constant of 11,000 N/m. The distance along the plane between m and M is 4.0 m. Mass m is released, makes an elastic collision with mass M, and rebounds a distance of 2.56 m back up the inclined plane. Mass M comes to rest momentarily 4.0 cm from its initial position. Find the mass M.

1. Use conservation of energy $mg\Delta h = \frac{1}{2}kx^2 - Mgx \sin 30°$; $\Delta h = 1.44 \sin 30° = 0.72$ m

2. Find M: $x = 0.04$ m, $m = 1$ kg, $k = 11\times10^4$ N/m $M = kx/g - 2m\Delta h/x$; $M = 8.85$ kg

---

129* •• The mass of a carbon nucleus is approximately 12 times the mass of a neutron. (a) Use the results of Problem 128 to show that after N head-on collisions of a neutron with carbon nuclei at rest, the energy of the neutron is approximately $0.716^N E_0$, where E_0 is its original energy. Neutrons emitted in the fission of a uranium nucleus have an energy of about 2 MeV. For such a neutron to cause the fission of another uranium nucleus in a reactor, its energy must be reduced to about 0.02 eV. (b) How many head-on collisions are needed to reduce the energy of a neutron from 2 MeV to 0.02 eV, assuming elastic head-on collisions with stationary carbon nuclei?

(a) 1. Write $K_{nf}/K_{ni} = (K_{ni} - \Delta K_n)/K_{ni}$ (see 8-128b) $K_{nf}/K_{ni} = (M - m)^2/(M + m)^2 =$ fractional loss per collision

    2. In this case $K_{nf}/K_{ni} = 0.716$ After N collisions, $K_{nf} = K_0 \times 0.716^N$

(b) In this case $(0.716)_N = 10$-8; solve for N $-8 = N \log(0.716)$; $N = 55$

133* •• Repeat Problem 24 if the cup has a mass m_c and the ball collides with it inelastically.

(a) The force exerted by the spring on $m_p = m_b g$ $F = kd + m_p g$

(b) 1. Find v_{bi}; use $p_i = p_f$ to find v_{cm} of ball + cup $v_{bi} = \sqrt{2gh}$; $v_{cm} = \sqrt{2gh}\,[m_b/(m_c + m_b)]$

    2. Apply energy conservation $kx^2 = (m_c+m_b)[m_b/(m_c+m_b)]^2(2gh) = 2m_b^2gh/(m_b+m_c)$

3. Solve for the compression x; then multiply by k to find the force the spring exerts on the platform.

$$x = m_b \sqrt{\frac{2gh}{k(m_c + m_b)}} \; ; \; kx = m_b g \sqrt{\frac{2kh}{g(m_c + m_b)}}$$

4. $F = m_p g + kx$

$$F = g\left(m_p + m_b \sqrt{\frac{2kh}{g(m_c + m_b)}} \right)$$

(c) Since the collision is inelastic, the ball never returns to its original position

---

137* •• A neutron at rest decays into a proton plus an electron. The conservation of momentum implies that the electron and proton should have equal and opposite momentum. However, experimentally they do not. This apparent nonconservation of momentum led Wolfgang Pauli to suggest in 1931 that there was a third, unseen particle emitted in the decay. This particle is called a neutrino, and it was finally observed directly in 1957. Suppose that the electron has momentum $p = 4.65 \times 10^{-22}$ kg.m/s along the negative x direction and the proton ($m = 1.67 \times 10^{-27}$ kg) moves with speed 2.93×10^5 m/s at an angle $17.9°$ above the x axis. Find the momentum of the neutrino. (The kinetic energy of the electron is comparable to its rest energy, so its energy and momentum are related relativistically rather than classically. However, the rest energy of the proton is large compared with its kinetic energy so the classical relation $E = \frac{1}{2}mv^2 = p^2/2m$ is valid.)

1. Momentum conservation: $\boldsymbol{p}_e + \boldsymbol{p}_p + \boldsymbol{p}_v = 0$

$p_p = 1.67 \times 10^{-27} \times 2.93 \times 10^5$ kg·m/s
$= 4.89 \times 10^{-22}$ kg·m/s

2. Since $p_{px} + p_e = 0$, $\boldsymbol{p}_v = -p_{py}\boldsymbol{j}$

$p_{px} = 4.89 \times 10^{-22}\cos 17.9° = 4.65 \times 10^{-22}$ kg·m/s $= -p_e$
$\boldsymbol{p}_v = -4.89 \times 10^{-22}\sin 17.9°\, \boldsymbol{j}$ kg·m/s
$= -1.5 \times 10^{-22}\, \boldsymbol{j}$ kg·m/s

CHAPTER 9

Rotation

1* • Two points are on a disk turning at constant angular velocity, one point on the rim and the other halfway between the rim and the axis. Which point moves the greater distance in a given time? Which turns through the greater angle? Which has the greater speed? The greater angular velocity? The greater tangential acceleration? The greater angular acceleration? The greater centripetal acceleration?

1. The point on the rim moves the greater distance.   2. Both turn through the same angle.   3. The point on the rim has the greater speed   4. Both have the same angular velocity.   5. Both have zero tangential acceleration. 6. Both have zero angular acceleration.   7. The point on the rim has the greater centripetal acceleration.

5* • A wheel starts from rest with constant angular acceleration of 2.6 rad/s^2. After 6 s, (*a*) What is its angular velocity? (*b*) Through what angle has the wheel turned? (*c*) How many revolutions has it made? (*d*) What is the speed and acceleration of a point 0.3 m from the axis of rotation?

(*a*) $\omega = \alpha t$

(*b*), (*c*) $\theta = \frac{1}{2}\alpha t^2$

(*d*) $v = \omega r$, $a_c = r\omega^2$, $a_t = r\alpha$, $a = (a_t^2 + a_c^2)^{1/2}$

$\omega = (2.6 \times 6)$ rad/s = 15.6 rad/s

$\theta = 46.8$ rad = 7.45 rev

$v = (15.6 \times 0.3)$ m/s = 4.68 m/s;

$a = [(0.3 \times 15.6^2)^2 + (0.3 \times 2.6)^2]^{1/2}$ m/s^2 = 73 m/s^2

9* • A Ferris wheel of radius 12 m rotates once in 27 s. (*a*) What is its angular velocity in radians per second? (*b*) What is the linear speed of a passenger? What is the centripetal acceleration of a passenger?

(*a*) $\omega = 2\pi/27$ rad/s = 0.233 rad/s.

(*b*) $v = r\omega = 12 \times 0.233$ /s = 2.8 m/s. $a_c = r\omega^2 = 12 \times 0.233^2$ m/s^2 = 0.65 m/s^2.

13* • A circular space station of radius 5.10 km is a long way from any star. Its rotational speed is controllable to some degree, and so the apparent gravity changes according to the tastes of those who make the decisions. Dave the Earthling puts in a request for artificial gravity of 9.8 m/s^2 at the circumference. His secret agenda is to give the Earthlings a home-gravity advantage in the upcoming interstellar basketball tournament. Dave's request would require an angular speed of (*a*) 4.4×10^{-2} rad/s. (*b*) 7.0×10^{-3} rad/s. (*c*) 0.28 rad/s. (*d*) −0.22 rad/s. (*e*) 1300 rad/s.

(*a*) Use $a_c = r\omega^2$ and solve for ω.

17* •   The moment of inertia of an object of mass M (a) is an intrinsic property of the object. (b) depends on the choice of axis of rotation. (c) Is proportional to M regardless of the choice of axis. (d) both (b) and (c) are correct.

(d)

21* •   A disk is free to rotate about an axis. A force applied a distance d from the axis causes an angular acceleration α. What angular acceleration is produced if the same force is applied a distance $2d$ from the axis? (a) α (b) 2α (c) $\alpha/2$ (d) 4α (e) $\alpha/4$

(b) $\alpha \propto \tau = F\ell$.

25* ••   A pendulum consisting of a string of length L attached to a bob of mass m swings in a vertical plane. When the string is at an angle θ to the vertical, (a) what is the tangential component of acceleration of the bob? (b) What is the torque exerted about the pivot point? (c) Show that $\tau = I\alpha$ with $a_t = L\alpha$ gives the same tangential acceleration as found in part (a).

(a) The pendulum and the forces acting on it are shown. The tangential force is $mg \sin \theta$. Therefore, the tangential acceleration is $a_t = g \sin \theta$.

(b) The tension causes no torque. The torque due to the weight about the pivot is $mgL \sin \theta$.

(c) Here $I = mL^2$; so $\alpha = mgL \sin \theta / mL^2 = g \sin \theta / L$, and $a_t = g \sin \theta$.

29* •   A tennis ball has a mass of 57 g and a diameter of 7 cm. Find the moment of inertia about its diameter. Assume that the ball is a thin spherical shell.

$I = (2/3)MR^2$ (see Table 9-1) $I = (2/3) \times 0.057 \times 0.035^2$ kg·m^2 = 4.66×10^{-5} kg·m^2

33* •   Use the parallel-axis theorem to find the moment of inertia of a solid sphere of mass M and radius R about an axis that is tangent to the sphere (Figure 9-39).

$I_{cm} = (2/5)MR^2$ (see Table 9-1); use Equ. 9-21 $I = (2/5)MR^2 + MR^2 = (7/5)MR^2$

37* ••   Tracey and Corey are doing intensive research on theoretical baton-twirling. Each is using "The Beast" as a model baton: two uniform spheres, each of mass 500 g and radius 5 cm, mounted at the ends of a 30-cm uniform rod of mass 60 g (Figure 9-40). Tracey and Corey want to calculate the moment of inertia of The Beast about an axis perpendicular to the rod and passing through its center. Corey uses the approximation that the two spheres can be treated as point particles that are 20 cm from the axis of rotation, and that the mass of the rod is negligible. Tracey, however, makes her calculations without approximations. (a) Compare the two results. (b) If the spheres retained the same mass but were hollow, would the rotational inertia increase or decrease? Justify your choice with a sentence or two. It is not necessary to calculate the new value of I.

(a) 1. Use point mass approximation for I_{app} $I_{app} = (2 \times 0.5 \times 0.2^2)$ kg·m^2 = 0.04 kg·m^2

    2. Use Table 9-1 and Equ. 9-21 to find I $I = [2(2/5)(0.5 \times 0.05^2) + I_{app} + (1/12)(0.06 \times 0.3^2)]$ kg·m^2

                                                      = 0.04145 kg·m^2; $I_{app}/I = 0.965$

(b) The rotational inertia would increase because I_{cm} of a hollow sphere $> I_{cm}$ of a solid sphere.

41* •••   The density of the earth is not quite uniform. It varies with the distance r from the center of the earth as $\rho = C(1.22 - r/R)$, where R is the radius of the earth and C is a constant. (a) Find C in terms of the total mass M and the radius R. (b) Find the moment of inertia of the earth. (See Problem 40.)

(a) $M = \int dm = \int_0^R 4\pi\rho r^2 \, dr = 4\pi C \int_0^R 1.22 r^2 \, dr - \frac{4\pi C}{R} \int_0^R r^3 \, dr = \frac{4\pi}{3} 1.22 \, CR^3 - \pi CR^3$. $C = 0.508 \, M/R^3$.

(b) $I = \int dI = \frac{8\pi}{3}\int_0^R \rho r^4 dr = \frac{8\pi \times 0.508 M}{3R^3}\left[\int_0^R 1.22 r^4 dr - \frac{1}{R}\int_0^R r^5 dr\right] = \frac{4.26M}{R^3}\left[\frac{1.22}{5}R^5 - \frac{1}{6}R^5\right] = 0.329 MR^2.$

45* ••• Use integration to determine the moment of inertia of a thin circular hoop of radius R and mass M for rotation about a diameter. Check your answer by referring to Table 9-1.

Here, $dm = \lambda R\, d\theta$, and $dI = z^2\, dm$, where $z = R\sin\theta$. Thus, $I = \lambda R^3 \int_{-\pi}^{\pi}\sin^2\theta\, d\theta = \lambda\pi R^3 = \frac{1}{2}MR^2$, in agreement with Table 9-1 for a hollow cylinder of length $L = 0$.

49* • Four 2-kg particles are located at the corners of a rectangle of sides 3 m and 2 m as shown in Figure 9-43. (a) Find the moment of inertia of this system about the z axis. (b) The system is set rotating about this axis with a kinetic energy of 124 J. Find the number of revolutions the system makes per minute.

(a) Use Equ. 9-2 $I = 2[2^2 + 3^2 + (2^2 + 3^2)]$ kg·m² = 52 kg·m²

(b) Find $\omega = (2K/I)^{1/2}$ $\omega = (2\times 124/52)^{1/2}$ rad/s = 2.18 rad/s = 20.9 rev/min

---

53* •• Calculate the kinetic energy of rotation of the earth, and compare it with the kinetic energy of motion of the earth's center of mass about the sun. Assume the earth to be a homogeneous sphere of mass 6.0×10^{24} kg and radius 6.4×10^6 m. The radius of the earth's orbit is 1.5×10^{11} m.

1. Find ω of the earth's rotation $\omega = 2\pi$ rad/day $= (2\pi/24\times 60\times 60)$ rad/s
$= 7.27\times 10^{-5}$ rad/s

2. Find K_{rot}; use Table 9-1 $K_{rot} = (\frac{1}{2}\times 0.4\times 6\times 10^{24}\times 6.4^2\times 10^{12}\times 7.27^2\times 10^{-10})$ J
$= 2.6\times 10^{29}$ J

3. Find K_{orb}; $I = M_E R_{orb}^2$; $\omega_{orb} = 2\pi/3.156\times 10^7$ rad/s $K_{orb} = (\frac{1}{2}\times 6\times 10^{24}\times 1.5^2\times 10^{22}\times 2^2\times 10^{-14})$ J
$K_{orb} \approx 10^4 K_{rot}$ $= 2.7\times 10^{33}$ J

57* •• You set out to design a car that uses the energy stored in a flywheel consisting of a uniform 100-kg cylinder of radius R. The flywheel must deliver an average of 2 MJ of mechanical energy per kilometer, with a maximum angular velocity of 400 rev/s. Find the least value of R such that the car can travel 300 km without the flywheel having to be recharged.

1. Find total energy $K = (2\times 10^6\times 300)$ J $= 6\times 10^8$ J $= \frac{1}{2}\times 50\times R^2\times \omega^2$

2. Solve for R with $\omega = 800\pi$ rad/s $R = \sqrt{24\times 10^6/(800\pi)^2}$ m $= 1.95$ m

---

61* •• For the system in Problem 60, find the linear acceleration of each block and the tension in the string.

1. Write the equations of motion for the three objects $4a = T_1$; $2a = 2g - T_2$; $0.08(T_2 - T_1) = \frac{1}{2}\times 0.6\times 0.08^2\alpha$

2. Use $\alpha = a/r$ and solve for a $T_2 - T_1 = 2g - 6a = 0.3a$; $a = 2g/6.3 = 3.11$ m/s²

3. Find T_1 (acting on 4 kg) and T_2 (acting on 2 kg). $T_1 = 12.44$ N; $T_2 = T_1 + 0.3a = 13.37$ N

65* •• A 1200-kg car is being unloaded by a winch. At the moment shown in Figure 9-46, the gearbox shaft of the winch breaks, and the car falls from rest. During the car's fall, there is no slipping between the (massless) rope, the pulley, and the winch drum. The moment of inertia of the winch drum is 320 kg·m² and that of the pulley is 4 kg·m². The radius of the winch drum is 0.80 m and that of the pulley is 0.30 m. Find the speed of the car as it hits the water.

1. Use energy conservation and $\omega = v/r$ $mgh = \frac{1}{2}mv^2 + \frac{1}{2}I_w\omega_w^2 + \frac{1}{2}I_p\omega_p^2 = \frac{1}{2}v^2(m + I_w/r_w^2 + I_p/r_p^2)$

2. Solve for and evaluate v $v = [2mgh/(m + I_w/r_w^2 + I_p/r_p^2)]^{1/2} = 8.2$ m/s

69* ••   Two objects are attached to ropes that are attached to wheels on a common axle as shown in Figure 9-50. The total moment of inertia of the two wheels is 40 kg·m². The radii of the wheels are $R_1 = 1.2$ m and $R_2 = 0.4$ m. (a) If $m_1 = 24$ kg, find m_2 such that there is no angular acceleration of the wheels. (b) If 12 kg is gently added to the top of m_1, find the angular acceleration of the wheels and the tensions in the ropes.

(a) Find τ_{net} and set equal to 0 $\qquad\qquad\qquad$ $\tau = m_1gR_1 - m_2gR_2 = 0$; $m_2 = m_1R_1/R_2 = 72$ kg

(b) 1. Write the equations of motion $\qquad\qquad$ $T_1 = m_1(g - R_1\alpha)$; $T_2 = m_2(g + R_2\alpha)$; $\alpha = (T_1R_1 - T_2R_2)/I$

$\quad$ 2. Solve for and find α with $m_1 = 36$ kg, $\qquad$ $\alpha = (m_1R_1 - m_2R_2)g/(m_1R_1{}^2 + m_2R_2{}^2 + I) = 1.37$ rad/s²

$\qquad$ $m_2 = 72$ kg

$\quad$ 3. Substitute $\alpha = 1.37$ rad/s² to find T_1 and T_2 $\qquad$ $T_1 = 294$ N; $T_2 = 745$ N

73* ••   A uniform cylinder of mass m_1 and radius R is pivoted on frictionless bearings. A massless string wrapped around the cylinder connects to a mass m_2, which is on a frictionless incline of angle θ as shown in Figure 9-52. The system is released from rest with m_2 a height h above the bottom of the incline. (a) What is the acceleration of m_2? (b) What is the tension in the string? (c) What is the total energy of the system when m_2 is at height h? (d) What is the total energy when m_2 is at the bottom of the incline and has a speed v? (e) What is the speed v? (f) Evaluate your answers for the extreme cases of $\theta = 0°$, $\theta = 90°$, and $m_1 = 0$.

(a) 1. Write the equations of motion $\qquad\qquad$ $m_2a = m_2g\sin\theta - T$; $\tau = RT = \frac{1}{2}m_1R^2\alpha$, $T = \frac{1}{2}m_1a$

$\quad$ 2. Solve for a $\qquad\qquad\qquad\qquad\qquad$ $a = (g\sin\theta)/(1 + m_1/2m_2)$

(b) Solve for T $\qquad\qquad\qquad\qquad\qquad$ $T = (\frac{1}{2}m_1g\sin\theta)/(1 + m_1/2m_2)$

(c) Take $U = 0$ at $h = 0$ $\qquad\qquad\qquad\qquad$ $E = K + U = m_2gh$

(d) This is a conservative system $\qquad\qquad\quad$ $E = m_2gh$

(e) $U = 0$; $E = K = \frac{1}{2}m_2v^2 + \frac{1}{2}I\omega^2$; $\omega = v/R$ $\quad$ $m_2gh = \frac{1}{2}(m_2 + \frac{1}{2}m_1)v^2$; $v = \sqrt{(2gh)/(1 + m_1/2m_2)}$

(f) 1. For $\theta = 0$ $\qquad\qquad\qquad\qquad\qquad$ $a = T = 0$

$\quad$ 2. For $\theta = 90°$ $\qquad\qquad\qquad\qquad\quad$ $a = g/(1 + m_1/2m_2)$; $T = \frac{1}{2}m_1a$; $v = \sqrt{(2gh)/(1 + m_1/2m_2)}$

$\quad$ 3. For $m_1 = 0$ $\qquad\qquad\qquad\qquad\qquad$ $a = g\sin\theta$, $T = 0$, $v = \sqrt{2gh}$

77* ••   A solid cylinder and a solid sphere have equal masses. Both roll without slipping on a horizontal surface. If their kinetic energies are the same, then (a) the translational speed of the cylinder is greater than that of the sphere. (b) the translational speed of the cylinder is less than that of the sphere. (c) the translational speeds of the two objects are the same. (d) (a), (b), or (c) could be correct depending on the radii of the objects.

$K_c = (3/4)mv_c{}^2$; $K_s = (7/10)mv_s{}^2$. If $K_c = K_{s,}$ then $v_c < v_s$. (b)

81* ••   A ball rolls without slipping along a horizontal plane. Show that the frictional force acting on the ball must be zero. *Hint:* Consider a possible direction for the action of the frictional force and what effects such a force would have on the velocity of the center of mass and on the angular velocity.

Let us assume that $f \neq 0$ and acts along the direction of motion. Now consider the acceleration of the center of mass and the angular acceleration about the point of contact with the plane. Since $F_{net} \neq 0$, $a_{cm} \neq 0$. However, $\tau = 0$ since $\ell = 0$, so $\alpha = 0$. But $\alpha = 0$ is not consistent with $a_{cm} \neq 0$. Consequently, $f = 0$.

85* •   A hoop of radius 0.40 m and mass 0.6 kg is rolling without slipping at a speed of 15 m/s toward an incline of slope 30°. How far up the incline will the hoop roll, assuming that it rolls without slipping?

1. Find the energy at the bottom of the slope $\qquad\qquad$ $K = mv^2$

2. Use energy conservation; $mgL\sin 30° = K$ $\qquad\quad$ $L = 2v^2/g = 45.9$ m

89* ••   A hollow sphere and uniform sphere of the same mass m and radius R roll down an inclined plane from the same height H without slipping (Figure 9-55). Each is moving horizontally as it leaves the ramp. When the spheres hit the ground, the range of the hollow sphere is L. Find the range L' of the uniform sphere.

1. Find v of each object as it leaves ramp. Use $mgH = \tfrac{1}{2}mv_h^2 + \tfrac{1}{2}(2/3)mv_h^2; \; v_h^2 = 6gH/5$
   energy conservation. $mgH = \tfrac{1}{2}mv_u^2 + \tfrac{1}{2}(2/5)mv_u^2; \; v_u^2 = 10gH/7$

2. Since distance $\propto v$, $L'/L = v_u/v_h$ $L' = L(25/21)^{\frac{1}{2}} = 1.09L$

93* •••   A wheel has a thin 3.0-kg rim and four spokes each of mass 1.2 kg. Find the kinetic energy of the wheel when it rolls at 6 m/s on a horizontal surface.

1. Find I of the wheel $I = M_{rim}R^2 + 4[(1/3)M_{spoke}R^2]$

2. Write $K = K_{trans} + K_{rot}$; use $v = R\omega$ $K = \tfrac{1}{2}(7.8 + 3 + 1.6) \times 6^2 \text{ J} = 223 \text{ J}$

97* •••   (a) Find the angular acceleration of the cylinder in Problem 96. Is the cylinder rotating clockwise or counterclockwise? (b) What is the cylinder's linear acceleration relative to the table? Let the direction of $\boldsymbol{F}$ be the positive direction. (c) What is the linear acceleration of the cylinder relative to the block?
We begin by drawing the two free-body diagrams

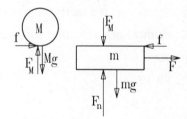

For the block, $F - f = ma_B$ (1). For the cylinder, $f = Ma_C$ (2).
Also, $fR = \tfrac{1}{2}MR^2\alpha$ and $f = \tfrac{1}{2}MR\alpha$. But $a_C = a_B - R\alpha$ or $R\alpha = a_B - a_C$.
Using Equs. (1) and (2) we now obtain $2f/M = a_B - f/M$ and
$3f/M = 3a_C = a_B$ (3). Equs. (1) and (3) yield $F - Ma_B/3 = ma_B$ and
solving for a_B one obtains $a_B = 3F/(M + 3m)$ and $a_C = F/(M + 3m)$.

(a) $\alpha = (a_B - a_C)/R = 2F/[R(M + 3m)]$. From the free-body diagram it is evident that the torque and, therefore, α is in the counterclockwise direction.

(b) The linear acceleration of the cylinder relative to the table is $a_C = F/(M + 3m)$.

(c) The acceleration of the cylinder relative to the block is $a_C - a_B = -2F/(M + 3m)$.

101*•   A cue ball is hit very near the top so that it starts to move with topspin. As it slides, the force of friction (a) increases v_{cm}. (b) decreases v_{cm}. (c) has no effect on v_{cm}.

(a)

105* ••   A uniform solid ball resting on a horizontal surface has a mass of 20 g and a radius of 5 cm. A sharp force is applied to the ball in a horizontal direction 9 cm above the horizontal surface. The force increases linearly from 0 to a peak value of 40,000 N in 10^{-4} s and then decreases linearly to 0 in 10^{-4} s. (a) What is the velocity of the ball after impact? (b) What is the angular velocity of the ball after impact? (c) What is the velocity of the ball when it begins to roll without sliding? For how long does the ball slide on the surface? Assume that $\mu_k = 0.5$.

(a) Find the translational impulse; then use $P_t = mv$ $F_{av} = 20,000 \text{ N}, \Delta t = 2 \times 10^{-4} \text{ s};$
    $v_0 = (4/0.02) \text{ m/s} = 200 \text{ m/s}$

(b) Find the rotational impulse about the CM $P_\tau = P_t(h - r) = I\omega_0$
    Solve for ω_0 with $I = (2/5)mr^2$ $\omega_0 = 5v_0(h - r)/2r^2$
    Evaluate ω_0 $\omega_0 = 5 \times 200 \times (0.09 - 0.05)/(2 \times 0.05^2) \text{ rad/s}$
    $= 8000 \text{ rad/s}$

(c) Note that $\omega_0 r = 400$ m/s $> v_0$; proceed as in $\omega = \omega_0 - (5/2)\mu_k gt/r$; $v = v_0 + \mu_k gt$; set $\omega r = v$; find t.

Example 9-16 $t = 2(\omega_0 r - v_0)/7\mu_k g = 11.6$ s; $v = 257$ m/s

---

109* ••   A solid cylinder of mass M resting on its side on a horizontal surface is given a sharp blow by a cue stick. The applied force is horizontal and passes through the center of the cylinder so that the cylinder begins translating with initial velocity v_0. The coefficient of sliding friction between the cylinder and surface is μ_k. (a) What is the translational velocity of the cylinder when it is rolling without slipping? (b) How far does the cylinder travel before it rolls without slipping? (c) What fraction of its initial mechanical energy is dissipated in friction?

This Problem is identical to Example 9-16 except that now $I = \frac{1}{2}MR^2$. Follow the same procedure.

(a) Set $\omega R = v$; $v = v_0 - \mu_k gt$; $\omega R = 2\mu_k gt$ $t = v_0/3\mu_k g$; $v = (2/3)v_0$

(b) $s = v_{av}t$ $s = 5v_0^2/18\mu_k g$

(c) $W_{fr}/K_i = (K_i - K_f)/K_i$ $K_i = \frac{1}{2}mv_0^2$; $K_f = (3/4)mv^2 = (1/3)mv_0^2$; $W_{fr}/K_i = 1/3$

113* ••   The radius of a park merry-go-round is 2.2 m. To start it rotating, you wrap a rope around it and pull with a force of 260 N for 12 s. During this time, the merry-go-round makes one complete rotation. (a) Find the angular acceleration of the merry-go-round. (b) What torque is exerted by the rope on the merry-go-round? (c) What is the moment of inertia of the merry-go-round?

(a) $\alpha = 2\theta/t^2$ $\alpha = 4\pi/12^2$ rad/s^2 = 0.0873 rad/s^2

(b) $\tau = Fr$ $\tau = (260 \times 2.2)$ N·m = 572 N·m

(c) $I = \tau/\alpha$ $I = (572/0.0873)$ kg·m^2 = 6552 kg·m^2

117* ••   A uniform rod of length L and mass m is pivoted at the middle as shown in Figure 9-64. It has a load of mass $2m$ attached to one of the ends. If the system is released from a horizontal position, what is the maximum velocity of the load?

1. Find I $I = mL^2/12 + 2mL^2/4 = 7mL^2/12$

2. $\frac{1}{2}I\omega^2 = 2mgL/2$; $v = \omega L/2$; solve for v $v = (2mgL/I)^{\frac{1}{2}}(L/2) = (6gL/7)^{\frac{1}{2}}$

121* ••   A hoop of mass 1.5 kg and radius 65 cm has a string wrapped around its circumference and lies flat on a horizontal frictionless table. The string is pulled with a force of 5 N. (a) How far does the center of the hoop travel in 3 s? (b) What is the angular velocity of the hoop about its center of mass after 3 s?

(a) $F_{net} = F = ma_{cm}$; $s = \frac{1}{2}a_{cm}t^2 = Ft^2/2m$ $s = (5 \times 3^2/2 \times 1.5)$ m = 15 m

(b) $\alpha = \tau/I$; $\omega = \alpha t = FRt/mR^2 = Ft/mR$ $\omega = (5 \times 3/1.5 \times 0.65)$ rad/s = 15.4 rad/s

---

125* ••   A spool of mass M rests on an inclined plane at a distance D from the bottom. The ends of the spool have radius R, the center has radius r, and the moment of inertia of the spool about its axis is I. A long string of negligible mass is wound many times around the center of the spool. The other end of the string is fastened to a hook at the top of the inclined plane such that the string always pulls parallel to the slope as shown in Figure 9-68. (a) Suppose that initially the slope is so icy that there is *no* friction. How does the spool move as it slips down the slope? Use energy considerations to determine the speed of the center of mass of the spool when it reaches the bottom of the slope. Give your answer in terms of M, I, r, R, g, D, and θ. (b) Now suppose that the ice is gone and that when the spool is set up in the same way, there is enough friction to keep it from slipping on the slope. What is the direction and magnitude of the friction force in this case?

(a) The spool will move down the plane at constant acceleration, spinning in a counterclockwise direction as string unwinds. From energy conservation, $MgD \sin \theta = \frac{1}{2}Mv^2 + \frac{1}{2}I\omega^2$; $v = r\omega$.

$$v = \sqrt{\frac{2MgD \sin \theta}{M + I/r^2}}.$$

(b) 1. The direction of the friction force is up along the plane

　2. Since $a_{cm} = 0$ and $\alpha = 0$, $F_{net} = 0$ and $\tau = 0$　　　$Mg \sin \theta = T + f_s$; $Tr = f_s R$

　3. Solve for f_s　　　　　　　　　　　　　　　　　$f_s = (Mg \sin \theta)/(1 + R/r)$

129* ••　Figure 9-71 shows a hollow cylinder of length 1.8 m, mass 0.8 kg, and radius 0.2 m. The cylinder is free to rotate about a vertical axis that passes through its center and is perpendicular to the cylinder's axis. Inside the cylinder are two masses of 0.2 kg each, attached to springs of spring constant k and unstretched lengths 0.4 m. The inside walls of the cylinder are frictionless. (a) Determine the value of the spring constant if the masses are located 0.8 m from the center of the cylinder when the cylinder rotates at 24 rad/s. (b) How much work was needed to bring the system from $\omega = 0$ to $\omega = 24$ rad/s?

Let $m = 0.2$ kg mass, $M = 0.8$ kg mass of cylinder, $L = 1.8$ m, and $x = $ distance of m from center $= x_0 + \Delta x$.

(a) We have $k\Delta x = m(x_0 + \Delta x)\omega^2$; solve for k　　　　$k = (0.2 \times 0.8 \times 24^2/0.4)$ N/m $= 230.4$ N/m

(b) $K = K_{rot} + \frac{1}{2}k\Delta x^2$; determine I of system when　　$I_M = \frac{1}{2}Mr^2 + ML^2/12 = 0.232$ kg·m^2

　　$x = 0.8$ m　　　　　　　　　　　　　　　　$I_{2m} = 2(mr^2/4 + mx^2) = 0.13$ kg·m^2; $I = 0.362$ kg·m^2

　　Evaluate $K = \frac{1}{2}I\omega^2 + \frac{1}{2}k\Delta x^2 = W$　　　　$W = (\frac{1}{2} \times 0.362 \times 24^2 + \frac{1}{2} \times 230.4 \times 0.4^2)$ J $= 122.7$ J

133* •••　A heavy, uniform cylinder has a mass m and a radius R (Figure 9-73). It is accelerated by a force T, which is applied through a rope wound around a light drum of radius r that is attached to the cylinder. The coefficient of static friction is sufficient for the cylinder to roll without slipping. (a) Find the frictional force. (b) Find the acceleration a of the center of the cylinder. (c) Is it possible to choose r so that a is greater than T/m? How? (d) What is the direction of the frictional force in the circumstances of part (c)?

(a) 1. Write the equations for translation and rotation　$T + f = ma$　　　　　　　　　(1)

　　　　　　　　　　　　　　　　　　　　　　$Tr - fR = I\alpha = \frac{1}{2}mRa$　　　　(2)

　2. Solve (2) for f　　　　　　　　　　　　$f = Tr/R - \frac{1}{2}ma$　　　　　(3)

　3. Use (3) in (1) to find a　　　　　　　　$a = (2T/3m)(1 + r/R)$　　　(4)

　4. Use (4) in (3) to find f in terms of T, r, and R　$f = (T/3)(2r/R - 1)$　　　(5)

(b) See Equ. (4) above　　　　　　　　　　Note: for $r = R$, results agree with Problem 9-131b

(c) Find r so that $a > T/m$　　　　　　　　From Equ. (4) above, $a > T/m$ if $r > \frac{1}{2}R$

(d) If $r > \frac{1}{2}R$ then $f > 0$, i.e., in the direction of T

CHAPTER 10

Conservation of Angular Momentum

1* • True or false: (*a*) If two vectors are parallel, their cross product must be zero. (*b*) When a disk rotates about its symmetry axis, ω is along the axis. (*c*) The torque exerted by a force is always perpendicular to the force.

(*a*) True (*b*) True (*c*) True

5* • Find $A \times B$ for (*a*) $A = 4\,i$ and $B = 6\,i + 6\,j$, (*b*) $A = 4\,i$ and $B = 6\,i + 6\,k$, and (*c*) $A = 2\,i + 3\,j$ and $B = -3\,i + 2\,j$.

Use Equ. 10-7; Note that $i \times i = j \times j = k \times k = 0$

(*a*) $A \times B = 24\,i \times j = 24\,k$. (*b*) $A \times B = 24\,i \times k = -24\,j$. (*c*) $A \times B = 4\,i \times j - 9\,j \times i = 13\,k$.

9* •• If $A = 3\,j$, $A \times B = 9\,i$, and $A \cdot B = 12$, find B.

Let $B = B_x\,i + B_y\,j + B_z\,k$; write $A \cdot B$ and find B_y $A \cdot B = 3B_y = 12$; $B_y = 4$

Write $A \times B$ and determine B_x and B_y $9\,i = 3B_x\,j \times i + 3B_z\,j \times k = -3B_x\,k + 3B_z\,i$; $B_x = 0$, $B_z = 3$

                                      $B = 4\,j + 3\,k$

13* •• A particle moves along a straight line at constant speed. How does its angular momentum about any point vary over time?

$L = mr \times p$ is constant.

17* •• A particle is traveling with a constant velocity v along a line that is a distance b from the origin O (Figure 10-30). Let dA be the area swept out by the position vector from O to the particle in time dt. Show that dA/dt is constant in time and equal to $\frac{1}{2}L/m$, where L is the angular momentum of the particle about the origin.

The area at $t = t_1$ is $A_1 = \frac{1}{2}br_1 \cos\theta_1 = \frac{1}{2}bx_1$, where θ_1 is the angle between r_1 and v and x_1 is the component of r_1 in the direction of v. At $t = t_1 + dt$, $A = A_1 + dA = \frac{1}{2}b(x + dx) = \frac{1}{2}b(x + v\,dt)$. Thus, $dA/dt = \frac{1}{2}bv = $ constant. Note that $r \sin\theta = b$; consequently, $\frac{1}{2}bv = \frac{1}{2}L/m$.

21* • A 1.8-kg particle moves in a circle of radius 3.4 m. The magnitude of its angular momentum relative to the center of the circle depends on time according to $L = (4\ \text{N·m})t$. (*a*) Find the magnitude of the torque acting on the particle. (*b*) Find the angular speed of the particle as a function of time.

(*a*) $\tau = dL/dt$ $\tau = 4\ \text{N·m}$

(*b*) $\omega = \alpha t$; $\alpha = \tau/I$; $I = mr^2$; $\omega = \tau t/mr^2$ $\omega = (4/1.8 \times 3.4^2)t\ \text{rad/s} = 0.192t\ \text{rad/s}$

25* •• Work Problem 24 for the case in which the coefficient of friction between the table and the 4-kg amplifier is 0.25.

Let m_1 be the speaker and m_2 be the amplifier. The forces acting on the system in the direction of motion are $m_1 g$ and $-m_2 g \mu_k$.

(a) Find the net torque about the pulley $\tau = gr(m_1 - m_2\mu_k) = 0.785$ N·m

(b) $L = \int \tau \, dt$ $L = 2.75$ kg·m^2/s

(c) $dL/dt = ar/(I_p/r^2 + m_1 + m_2) = gr(m_1 - m_2\mu_k)$ $a = g(m_1 - m_2\mu_k)/(I_p/r^2 + m_1 + m_2) = 1.56$ m/s^2

    Find v at $t = 3.5$ s; $v = at$. $\omega = v/r$ $v = 5.45$ m/s; $\omega = (5.45/0.08)$ rad/s $= 68.1$ rad/s

    $L_p = I_p\omega$; $I_p = \frac{1}{2}m_p r^2$ $L_p = (0.3 \times 0.08^2 \times 68.1)$ kg·m^2/s $= 0.131$ kg·m^2/s

(d) Find $L(m_1)$ and $L(m_2)$; $L = mrv$ $L(m_1) = (2 \times 0.08 \times 5.45)$ kg·m^2/s $= 0.872$ kg·m^2/s

                                      $L(m_2) = (4 \times 0.08 \times 5.45)$ kg·m^2/s $= 1.744$ kg·m^2/s

    Evaluate the ratios $L(m_1)/L_{tot} = 0.317$; $L(m_2)/L_{tot} = 0.634$; $L_p/L_{tot} = 0.048$

29* • If the angular momentum of a system is constant, which of the following statements must be true? (a) No torque acts on any part of the system. (b) A constant torque acts on each part of the system. (c) Zero net torque acts on each part of the system. (d) A constant external torque acts on the system. (e) Zero net torque acts on the system.

(e)

33* •• A block sliding on a frictionless table is attached to a string that passes through a hole in the table. Initially, the block is sliding with speed v_0 in a circle of radius r_0. A student under the table pulls slowly on the string. What happens as the block spirals inward? Give supporting arguments for your choice. (a) Its energy and angular momentum are conserved. (b) Its angular momentum is conserved, and its energy increases. (c) Its angular momentum is conserved, and its energy decreases. (d) Its energy is conserved, and its angular momentum increases. (e) Its energy is conserved, and its angular momentum decreases.

(b) $\tau = 0$, so L is conserved. The student does work, $Fs \neq 0$, so the energy of the block must increase.

37* •• A small blob of putty of mass m falls from the ceiling and lands on the outer rim of a turntable of radius R and moment of inertia I_0 that is rotating freely with angular speed ω_i about its vertical fixed symmetry axis. (a) What is the postcollision angular speed of the turntable plus putty? (b) After several turns, the blob flies off the edge of the turntable. What is the angular speed of the turntable after the blob flies off?

(a) $\tau_{ext} = 0$; $I_0\omega_i = (I_0 + mR^2)\omega_f$ $\omega_f = \omega_i/(1 + mR^2/I_0)$

(b) When m flies off, its angular momentum does $\omega' = \omega_f$
    not change

41* •• The sun's radius is 6.96×10^8 m, and it rotates with a period of 25.3 d. Estimate the new period of rotation of the sun if it collapses with no loss of mass to become a neutron star of radius 5 km.

$\omega_2 = \omega_1(R_1^2/R_2^2)$ (see Problem 35); $T_2 = T_1(R_2/R_1)^2$; $T_2 = [25.3(5/6.96 \times 10^5)^2]$ days $= 1.31 \times 10^{-9}$ days $= 0.11$ ms. Note: This assumes that the mass distribution in the sun and neutron star are the same. However, the sun's mass is concentrated near its center, whereas the density of the neutron star is nearly constant. The correct period will be substantially greater than 0.11 ms.

45* • The z component of the spin of an electron is $\frac{1}{2}\hbar$, but the magnitude of the spin vector is $\sqrt{0.75}\,\hbar$. What is the angle between the electron's spin angular momentum vector and the z axis?

$\cos\theta = 0.5/(0.75)^{\frac{1}{2}}$; $\theta = 54.7°$.

49* •• A 16.0-kg, 2.4-m-long rod is supported on a knife edge at its midpoint. A 3.2-kg ball of clay is dropped from rest from a height of 1.2 m and makes a perfectly inelastic collision with the rod 0.9 m from the point of support (Figure 10-38). Find the angular momentum of the rod-and-clay system immediately after the inelastic collision.

$L_i = L_f$; find L just prior to collision $L = mvr = [3.2 \times (2 \times 9.81 \times 1.2)^{\frac{1}{2}} \times 0.9]$ J·s = 14 J·s

53* •• If, for the system of Problem 52, $L = 1.2$ m, $M = 0.8$ kg, and $m = 0.3$ kg, and the maximum angle between the rod and the vertical is 60°, find the speed of the particle before impact.

We proceed as follows: Use conservation of L about the pivot to find ω immediately after collision, then use energy conservation to determine v of mass m before collision for an arbitrary angle θ. Then set $\theta = 60°$.

1. Conservation of angular momentum: $0.8Lmv = I\omega = (ML^2/3 + 0.64L^2m)\omega$. $\omega = (0.8Lmv)/(ML^2/3 + 0.64mL^2)$.

2. Conservation of energy: $[MgL/2 + mg(0.8L)](1 - \cos\theta) = \frac{1}{2}I\omega^2 = \dfrac{0.32(Lmv)^2}{ML^2/3 + 0.64\,mL^2}$

3. Solve for v: $v = \sqrt{\dfrac{(0.5M + 0.8m)(ML^2/3 + 0.64mL^2)g(1 - \cos\theta)}{0.32Lm^2}}$

4. Substitute numerical values for L, M, m, and θ. $v = 7.75$ m/s.

57* •• Suppose that in Figure 10-42, $m = 0.4$ kg, $M = 0.75$ kg , $L_1 = 1.2$ m, and $L_2 = 0.8$ m. What minimum initial angular velocity must be imparted to the rod so that the system will revolve completely about the hinge following the inelastic collision? How much energy is then dissipated in the inelastic collision?

Let ω_i and ω_f be the angular velocities of the rod immediately before and immediately after the inelastic collision with the mass m. Let ω_0 be the initial angular velocity of the rod. We proceed as follows: 1. We apply energy conservation to determine ω_f. 2. Next we apply conservation of angular momentum to determine ω_i. 3. Next, we again apply energy conservation to determine ω_0. 4. Finally, we find the energies of the system immediately before and immediately after the collision and, thereby, the energy loss.

1. Set K immediately after collision equal to potential energy after 180° rotation.

  $\frac{1}{2}(ML_1^2/3 + mL_2^2)\omega_f^2 = MgL_1 + 2mgL_2$; evaluate ω_f: $\omega_f = 7.0$ rad/s.

2. $(ML_1^2/3)\omega_i = (ML_1^2/3 + mL_2^2)\omega_f$; solve for and evaluate ω_i; $\omega_i = [(0.96 + 0.256)/0.96]\omega_f = 8.87$ rad/s.

3. $\frac{1}{2}(ML_1^2/3)\omega_i^2 = \frac{1}{2}(ML_1^2/3)\omega_0^2 + MgL_1/2$; $\omega_0^2 = \omega_i^2 - 3g/L_1$; $\omega_0 = 7.355$ rad/s.

4. $E_i = \frac{1}{2}(ML_1^2/3)\omega_i^2 = 40.95$ J; $E_f = MgL_1 + 2mgL_2 = 29.82$ J. $\Delta E = 11.13$ J.

61* •• A man is walking north carrying a suitcase that contains a spinning gyroscope mounted on an axle attached to the front and back of the case. The angular velocity of the gyroscope points north. The man now begins to turn to walk east. As a result, the front end of the suitcase will (*a*) resist his attempt to turn and will try to remain pointed north. (*b*) fight his attempt to turn and will pull to the west. (*c*) rise upward. (*d*) dip downward. (*e*) cause no effect whatsoever.

(*d*)

65* •• A uniform disk of mass 2.5 kg and radius 6.4 cm is mounted in the center of a 10-cm axle and spun at 700 rev/min. The axle is then placed in a horizontal position with one end resting on a pivot. The other end is given an initial horizontal velocity such that the precession is smooth with no nutation. (a) What is the angular velocity of precession? (b) What is the speed of the center of mass during the precession? (c) What are the magnitude and direction of the acceleration of the center of mass? (d) What are the vertical and horizontal components of the force exerted by the pivot?

(a) $\omega_p = MgD/I_s\omega_s$; $I_s = \frac{1}{2}MR^2$; $\omega_p = 2gD/R^2\omega_s$
$\quad\quad \omega_p = [2 \times 9.81 \times 0.05/(0.064^2 \times 700 \times 2\pi/60)]$
$\quad\quad\quad = 3.27$ rad/s

(b) $v_{cm} = \omega_p D$
$\quad\quad v_{cm} = 3.27 \times 0.05$ m/s $= 0.163$ m/s

(c) $a_{cm} = D\omega_p^2$
$\quad\quad a_{cm} = 0.535$ m/s^2

(d) $F_v = Mg$; $F_h = Ma_{cm}$
$\quad\quad F_v = 24.5$ N; $F_h = 1.34$ N

69* •• In tetherball, a ball is attached to a string that is attached to a pole. When the ball is hit, the string wraps around the pole and the ball spirals inward. Neglecting air resistance, what happens as the ball swings around the pole? Give supporting arguments for your choice. (a) The mechanical energy and angular momentum of the ball are conserved. (b) The angular momentum of the ball is conserved, but the mechanical energy of the ball increases. (c) The angular momentum of the ball is conserved, and the mechanical energy of the ball decreases. (d) The mechanical energy of the ball is conserved and the angular momentum of the ball increases. (e) The mechanical energy of the ball is conserved and the angular momentum of the ball decreases.

(e) Consider the situation shown in the adjoining figure. The ball rotates counterclockwise. The torque about the center of the pole is clockwise and of magnitude RT, where R is the pole's radius and T is the tension. So L must decrease.

73* •• An ice skater starts her pirouette with arms outstretched, rotating at 1.5 rev/s. Estimate her rotational speed (in revolutions per second) when she brings her arms flat against her body.
Assume I of body, minus arms, $= \frac{1}{2} \times 50 \times 0.2^2 = 1.0$ kg·m^2. The mass of each arm $= 4$ kg, and length $= 1.0$ m. Then I_{tot}(arms out) $= 1.0 + 2 \times 4/3 = 3.7$ kg·m^2 and I_{tot}(arms in) $= 1.0 + 2 \times 4 \times 0.2^2 = 1.32$ kg·m^2. With L constant, one then finds that $\omega = 1.5(3.7/1.32) = 4.2$ rev/s.

77* •• Figure 10-45 shows a hollow cylindrical tube of mass M, length L, and moment of inertia $ML^2/10$. Inside the cylinder are two masses m, separated a distance ℓ and tied to a central post by a thin string. The system can rotate about a vertical axis through the center of the cylinder. With the system rotating at ω, the strings holding the masses suddenly break. When the masses reach the end of the cylinder, they stick. Obtain expressions for the final angular velocity and the initial and final energies of the system. Assume that the inside walls of the cylinder are frictionless.

1. $\tau = 0$; $L_f = L_i$; $\omega_f = \omega(I_i/I_f)$. Obtain expressions for I_i and I_f and ω_f.

$\quad\quad I_i = (ML^2/10 + 2m\ell^2/4)$; $I_f = (ML^2/10 + 2mL^2/4)$ $\quad$ (1)

$\quad\quad \omega_f = [(M + 5m\ell^2/L^2)/(M + 5m)]\omega$ $\quad\quad\quad\quad\quad\quad$ (2)

2. $K_i = \frac{1}{2}I_i\omega_i^2$; $K_f = \frac{1}{2}I_f\omega_f^2$

$\quad\quad K_i = (ML^2 + 5m\ell^2)\omega^2/20$ $\quad\quad\quad\quad\quad\quad\quad\quad\quad$ (3)

$\quad\quad K_f = [(ML^2 + 5m\ell^2)^2/(ML^2 + 5mL^2)]\omega^2/20$ $\quad\quad$ (4)

81* •• · Given the numerical values of Problem 79, suppose the coefficient of friction between the masses and the walls of the cylinder is such that the masses cease sliding 0.2 m from the ends of the cylinder. Determine the initial and final angular velocities of the system and the energy dissipated in friction.

1. Find ω_i, the initial angular velocity when the string breaks	$m(\ell/2)\omega^2 = T$; $\omega = \sqrt{2T/m\ell}$ $\omega_i = 30$ rad/s
2. Find I_i and I_f; see Problem 77 for I_i	$I_i = 0.392$ kg·m²; $I_f = ML^2/10 + 2m \times 0.8^2 = 0.832$ kg·m²
3. Use $L_i = L_f$; $\omega_f = \omega_i(I_i/I_f)$	$\omega_f = 14.13$ rad/s
4. Find K_i, K_f and $\Delta E = K_i - K_f$; $K = \frac{1}{2}I\omega^2$	$K_i = 176.4$ J; $K_f = 83.1$ J; $\Delta E = 93.3$ J

85* •• The polar ice caps contain about 2.3×10^{19} kg of ice. This mass contributes negligibly to the moment of inertia of the earth because it is located at the poles, close to the axis of rotation. Estimate the change in the length of the day that would be expected if the polar ice caps were to melt and the water were distributed uniformly over the surface of the earth. (The moment of inertia of a spherical shell of mass m and radius r is $2mr^2/3$.)

1. $T = 2\pi I/L$ (see Problem 84); L is constant	$dT/T = dI/I$ or $\Delta T = T\Delta I/I$
2. $I = 2M_E R_E^2/5$; $M_E = 6 \times 10^{24}$ kg; $\Delta I = 2mR_E^2/3$	$\Delta T = (1\ d)[(5/3)(2.3 \times 10^{19}/6 \times 20^{24})] = 6.4 \times 10^{-6}\ d$ $\doteq 0.55$ s

---

89* ••• Figure 10-48 shows a pulley in the shape of a uniform disk with a heavy rope hanging over it. The circumference of the pulley is 1.2 m and its mass is 2.2 kg. The rope is 8.0 m long and its mass is 4.8 kg. At the instant shown in the figure, the system is at rest and the difference in height of the two ends of the rope is 0.6 m. (a) What is the angular velocity of the pulley when the difference in height between the two ends of the rope is 7.2 m? (b) Obtain an expression for the angular momentum of the system as a function of time while neither end of the rope is above the center of the pulley. There is no slippage between rope and pulley.

(a) Take $y = 0$ at center of pulley; write U_i and U_f for the free part of the rope. Let $\lambda = M_r/L$ = 0.6 kg/m.	$U_i = -\frac{1}{2}L_{1i}(L_{1i}\lambda g) - \frac{1}{2}L_{2i}(L_{2i}\lambda g) = -\frac{1}{2}(L_{1i}^2 + L_{2i}^2)\lambda g$ $U_f = -\frac{1}{2}(L_{1f}^2 + L_{2f}^2)\lambda g$; $L_{1i}+L_{2i} = 7.4$ m; $L_{2i}-L_{2i} = 0.6$ m;
Let L_1 and L_2 be the lengths of the hanging parts.	$L_{1i} = 3.4$ m, $L_{2i} = 4.0$ m; also $L_{1f} = 0.1$ m, $L_{2f} = 7.3$ m
Use conservation of energy: $K + \Delta U = 0$.	$K = \frac{1}{2}I_p\omega^2 + \frac{1}{2}Mv^2$; $v = R\omega$, $R = (1.2/2\pi)$ m $= 0.6/\pi$ m
Write K in terms of ω.	$K = \frac{1}{2} \times \frac{1}{2} \times 2.2(0.6/\pi)^2\omega^2 + \frac{1}{2} \times 8(0.6/\pi)^2\omega^2 = 0.166\omega^2$
Find ΔU	$\Delta U = U_f - U_i = -\frac{1}{2}(0.1^2 - 3.4^2 + 7.3^2 - 4.0^2) \times 0.6g$ $= -75.75$ J
Solve for ω	$\omega = (75.75/0.166)^{\frac{1}{2}}$ rad/s $= 21.36$ rad/s
(b) Now write $U(\theta)$ and ΔU, where θ is the angle through which the pulley has turned. This will reduce L_1 by $R\theta$ and increase L_2 by $R\theta$.	$U(\theta) = -\frac{1}{2}[(L_{1i} - R\theta)^2 + (L_{2i} + R\theta)^2]\lambda g$ $\Delta U = -U(\theta) + \frac{1}{2}(L_{1i}^2 + L_{2i}^2)\lambda g = R^2\theta^2\lambda g$
Use energy conservation and result for K	$0.166\omega^2 = [(0.6/\pi)^2 \times 0.6 \times 9.81]\theta^2 = 0.215\theta^2$; $\omega = 1.14\theta$
Recall that $\omega = d\theta/dt$	$d\theta/dt = 1.14\theta$; $d\theta/\theta = 1.14\ dt$
Integrate	$\ln(\theta) = 1.14t$; $\theta(t) = e^{1.14t} - 1$

Find $\omega(t)$

$\omega(t) = d\theta/dt = 1.14e^{1.14t}$

$L = L_p + L_r = (I_p + M_rR^2)\omega$; note that the angular momentum of each portion of the rope is the same

$L = [\frac{1}{2} \times 2.2(0.6/\pi)^2 + 8.0(0.6/\pi)^2] \times 1.14e^{1.14t}$ J·s

$L = 0.378e^{1.14t}$ J·s

CHAPTER **11**

Gravity

1* • True or false: (*a*) Kepler's law of equal areas implies that gravity varies inversely with the square of the distance. (*b*) The planet closest to the sun, on the average, has the shortest period of revolution about the sun.
(*a*) False   (*b*) True

5* • A comet has a period estimated to be about 4210 y. What is its mean distance from the sun? (4210 y was the estimated period of the comet Hale–Bopp, which was seen in the Northern Hemisphere in early 1997. Gravitational interactions with the major planets that occurred during this apparition of the comet greatly changed its period, which is now expected to be about 2380 y.)
Use Equ. 11-2; $R_{mean} = (1\ AU)(4210)^{2/3}$ $R_{mean} = 1.5 \times 10^{11} \times 4210^{2/3}$ m $= 3.91 \times 10^{13}$ m

---

9* • Why don't you feel the gravitational attraction of a large building when you walk near it?
The mass of the building is insignificant compared to the mass of the earth.

13* • One of Jupiter's moons, Io, has a mean orbital radius of 4.22×10^8 m and a period of 1.53×10^5 s. (*a*) Find the mean orbital radius of another of Jupiter's moons, Callisto, whose period is 1.44×10^6 s. (*b*) Use the known value of *G* to compute the mass of Jupiter.
(*a*) Use Equ. 11-2; $R_C = R_I (T_C/T_I)^{2/3}$ $R_C = (4.22 \times 10^8)(14.4/1.53)^{2/3}$ m $= 18.8 \times 10^8$ m
(*b*) Use Equ. 11-15; $M_J = 4\pi^2 R_I^3/GT_I^2$ $M_J = 4\pi^2(4.22 \times 10^8)^3/[6.67 \times 10^{-11}(1.53 \times 10^5)^2]$
                                                                                  $= 1.9 \times 10^{27}$ kg

---

17* • An object is dropped from a height of 6.37×10^6 m above the surface of the earth. What is its initial acceleration?
Since $R = 2R_E$, $a = g_E/4 = (9.81/4)$ m/s$^2 = 2.45$ m/s^2.

21* • A comet orbits the sun with constant angular momentum. It has a maximum radius of 150 AU, and at aphelion its speed is 7×10^3 m/s. The comet's closest approach to the sun is 0.4 AU. What is its speed at perihelion?
$L = $ constant $= mv_p r_p = mv_a r_a$; $v_p = v_a r_a/r_p$ $v_p = 7 \times 10^3 \times 150/0.4$ m/s $= 2625$ km/s

---

25* •• Suppose that Kepler had found that the period of a planet's circular orbit is proportional to the square of the orbit radius. What conclusion would Newton have drawn concerning the dependence of the gravitational attraction on distance between two masses?

Take $F = CR^n$, where C is a constant. Then, for a stable circular orbit, $v^2/R = F = CR^n$. The period of the orbit is given by $T = 2\pi R/v$, and so $T = 2\pi R/C^{\frac{1}{2}}R^{(n+1)/2}$. Therefore, if $T \propto R^2$, $1 - (n + 1)/2 = 2$, $n = -3$, and $F \propto 1/R^3$.

29* •• The mass of the earth is 5.98×10^{24} kg and its radius is 6370 km. The radius of the moon is 1738 km. The acceleration of gravity at the surface of the moon is 1.62 m/s². What is the ratio of the average density of the moon to that of the earth?

1. Write g_E and g_M in terms of ρ_E and ρ_M

2. Find g_M/g_E and solve for and evaluate ρ_M/ρ_E

$g_E = G(4\pi\rho_E R_E^3/3)/R_E^2$; $g_M = G(4\pi\rho_M R_M^3/3)/R_M^2$

$g_M/g_E = \rho_M R_M/\rho_E R_E$; $\rho_M/\rho_E = (1.62/9.81)(6.37/1.738)$

$= 0.605$

33* • The masses in a Cavendish apparatus are $m_1 = 12$ kg and $m_2 = 15$ g, and the separation of their centers is 7 cm. (a) What is the force of attraction between these two masses? (b) If the rod separating the two small masses is 18 cm long, what torque must be exerted by the suspension to balance the torque exerted by gravity?

(a) Use Equs. 11-3 and 11-4

$F = (6.67 \times 10^{-11} \times 12 \times 1.5 \times 10^{-2}/49 \times 10^{-4})$ N

$= 2.45 \times 10^{-9}$ N

(b) $\tau = 2Fr$, $r = 0.09$ m

$\tau = 4.41 \times 10^{-10}$ N·m

37* • The weight of a standard object defined as having a mass of exactly 1 kg is measured to be 9.81 N. In the same laboratory, a second object weighs 56.6 N. (a) What is the mass of the second object? (b) Is the mass you determined in part (a) gravitational or inertial mass?

(a) $m_2 = (56.6/9.81)$ kg = 5.77 kg. (b) This is the gravitational mass of m_2—determined by the effect on m_2 of the earth's gravitational field.

41* •• An object is dropped from rest from a height of 4×10^6 m above the surface of the earth. If there is no air resistance, what is its speed when it strikes the earth?

1. Use $\frac{1}{2}mv^2 = -\Delta U = U(4 \times 10^6 + R_E) - U(R_E)$

2. Solve for v

$\frac{1}{2}v^2 = (3.99 \times 10^{14})[(1/6.37 \times 10^6) - (1/10.37 \times 10^6)]$ J/kg

$v = 6.95$ km/s

45* ••• The assumption of uniform mass density in Problem 44 is rather unrealistic. For most galaxies, the mass density increases greatly toward the center of the galaxy. Repeat Problem 44 using a surface mass density of the form $\sigma(r) = C/r$, where $\sigma(r)$ is the mass per unit area of the disk at a distance r from the center. First determine the constant C in terms of R and M; then proceed as in Problem 44.

(a) $M = \int_0^R 2\pi r\sigma \, dr = 2\pi C \int_0^R dr = 2\pi CR$

$C = M/2\pi R$

(b) Write an expression for dU

$dU = -\dfrac{GM \, dr}{R\sqrt{x^2 + r^2}}$

$U(x) = \int_0^R dU$

$U(x) = -\dfrac{GM}{R} \int_0^R \dfrac{dr}{\sqrt{x^2 + r^2}} = -\dfrac{GM}{R} \ln\left(\dfrac{R + \sqrt{x^2 + R^2}}{x}\right)$

(c) $F_x = -dU/dx = g(x)$ $F_x = \dfrac{GM}{x\sqrt{x^2 + R^2}} = g(x)$

49* •   The planet Saturn has a mass 95.2 times that of the earth and a radius 9.47 times that of the earth. Find the escape speed for objects near the surface of Saturn.

$v_e(Sat) = (2GM_S/R_S)^{1/2} = v_e(earth)[(M_S/M_E)(R_E/R_S)]^{1/2}$ $v_e(Sat) = 11.2(95.2/9.47)^{1/2}$ km/s $= 35.5$ km/s

---

53* ••   A space probe launched from the earth with an initial speed v_i is to have a speed of 60 km/s when it is very far from the earth. What is v_i?

$\frac{1}{2}mv_i^2 = \frac{1}{2}mv_f^2 + \frac{1}{2}mv_e^2; \ v_i = (v_f^2 + v_e^2)^{1/2}$ $v_i = (60^2 + 11.2^2)^{1/2}$ km/s $= 61.04$ km/s

---

57* ••   If an object has just enough energy to escape from the earth, it will not escape from the solar system because of the attraction of the sun. Use Equation 11-19 with M_S replacing M_E and the distance to the sun r_S replacing R_E to calculate the speed v_{es} needed to escape from the sun's gravitational field for an object at the surface of the earth. Neglect the attraction of the earth. Compare your answer with that in Problem 56. Show that if v_e is the speed needed to escape from the earth, neglecting the sun, then the speed of an object at the earth's surface needed to escape from the solar system is given by $v_{e,solar}^2 = v_e^2 + v_{es}^2$, and calculate $v_{e,solar}$.

From Equ. 11-19, $v_{es} = (2GM_S/r_S)^{1/2} = [2 \times (6.67 \times 10^{-11}) \times (2 \times 10^{30})/1.5 \times 10^{11}]^{1/2} = 42.2$ km/s; this is the same speed calculated in Problem 56, as it should be. The energy needed to escape the solar system, starting from the surface of the earth, is the sum of the energy needed to escape the earth's gravity plus that required to escape from the sun, starting at the earth's orbit radius. These energies are proportional to the squares of the corresponding escape velocities. Therefore, $v_{e,solar}^2 = v_e^2 + v_{es}^2; \ v_{e,solar} = (11.2^2 + 42.2^2)^{1/2}$ km/s $= 43.7$ km/s.

61* ••   A spacecraft of 100 kg mass is in a circular orbit about the earth at a height $h = 2R_E$. (a) What is the period of the spacecraft's orbit about the earth? (b) What is the spacecraft's kinetic energy? (c) Express the angular momentum L of the spacecraft about the earth in terms of its kinetic energy K and find its numerical value. We will use the result of Problem 55, i.e., $v_c = v_e/\sqrt{2}$, where v_c is the speed of a circular orbit just above the surface.

(a) 1. $K(3R_E) = K(R_E)/3; \ v = v_c/\sqrt{3} = v_e/\sqrt{6}$ $v = 11.2/\sqrt{6}$ km/s $= 4.57$ km/s

   2. $T = 2\pi(3R_E)/v$ $T = 2\pi \times 3 \times 6370/(4.57 \times 3600)$ h $= 7.3$ h

(b) $K = \frac{1}{2}mv^2$ $K = 50 \times (4.57 \times 10^3)^2 = 1.04 \times 10^9$ J

(c) $K = L^2/2I; \ L = (2KI)^{1/2}; \ I = m(3R_E)^2$ $L = 3R_E(2Km)^{1/2} = 3R_Emv = 8.73 \times 10^{10}$ J·s

---

65* •   The gravitational field at some point is given by $g = 2.5 \times 10^{-6}$ N/kg j. What is the gravitational force on a mass of 4 g at that point?

Use Equ. 11-21 $F = mg = (4 \times 10^{-3} \times 2.5 \times 10^{-6})j$ N $= 10^{-8}j$ N

69* ••   (a) Show that the gravitational field of a ring of uniform mass is zero at the center of the ring. (b) Figure 11-22 shows a point P in the plane of the ring but not at its center. Consider two elements of the ring of length s_1 and s_2 at distances of r_1 and r_2, respectively. 1. What is the ratio of the masses of these elements? 2. Which produces the greater gravitational field at point P? 3. What is the direction of the field at point P due to these elements? (c) What is the direction of the gravitational field at point P due to the entire ring? (d) Suppose that the gravitational field varied as $1/r$ rather than $1/r^2$. What would be the net gravitational field at point P due to the two elements? (e) How would your answers to parts (b) and (c) differ if point P were inside a spherical shell of uniform mass rather than inside a plane circular ring?

Let λ = mass per unit length of the ring.

(a) g of opposite elements of mass $R\lambda\,d\theta$ cancel   By symmetry $g = 0$ at center

(b) 1. $m_1 = r_1\lambda\,d\theta$; $m_2 = r_2\lambda\,d\theta$ $m_1/m_2 = r_1/r_2$

 2. $g = Gm/r^2 = Gr\lambda\,d\theta/r^2 = G\lambda\,d\theta/r$ $r_1 < r_2$, therefore $g_1 > g_2$

 3. By symmetry, g points along OP g points toward m_1, i.e., in direction of OP

(c) Take x along OP By symmetry, $g_y = 0$; g points in the direction OP

(d) For $g \propto 1/r$, $g_1 = g_2 \propto \lambda\,d\theta$ $g_1 = -g_2$; $g = 0$

(e) Now m_1 and $m_2 \propto r^2$, so $g_1 = g_2$ $g_1 = -g_2$; $g = 0$; note: $g = 0$ everywhere inside the shell

---

73* •• Explain why the gravitational field increases with r rather than decreasing as $1/r^2$ as one moves out from the center inside a solid sphere of uniform mass.

g is proportional to the mass within the sphere and inversely proportional to the radius, i.e., proportional to $r^3/r^2 = r$.

77* •• Two homogeneous spheres, S_1 and S_2, have equal masses but different radii, R_1 and R_2. If the acceleration of gravity on the surface of sphere S_1 is g_1, what is the acceleration of gravity on the surface of sphere S_2?

$g \propto M/r^2$ $g_2 = g_1(R_1^2/R_2^2)$

81* •• A sphere of radius R has its center at the origin. It has a uniform mass density ρ_0, except that there is a spherical cavity in it of radius $r = \frac{1}{2}R$ centered at $x = \frac{1}{2}R$ as in Figure 11-24. Find the gravitational field at points on the x axis for $|x| > R$. (*Hint:* The cavity may be thought of as a sphere of mass $m = (4/3)\pi r^3\rho_0$ plus a sphere of mass $-m$.)

Write $g(x)$ using the hint. That is, find the sum of g of the solid sphere plus the field of a sphere of radius $\frac{1}{2}R$ of negative mass centered at $x = \frac{1}{2}R$.

$$g(x) = G\left(\frac{4\pi\rho_0}{3}\right)\left(\frac{R^3}{x^2} - \frac{R^3/8}{(x - \frac{1}{2}R)^2}\right)$$

85* ••• A hole is drilled into the sphere of Problem 84 toward the center of the sphere to a depth of 2 km below the sphere's surface. A small mass is dropped from the surface into the hole. Determine the speed of the small mass as it strikes the bottom of the hole.

1. Write the work per kg done by g between $r = 5$ m and $r = 3$ m; note that g points inward

$$E = \int_5^3 g\,dr = 2\pi GC(5-3) = 5.4 \times 10^{-9}\ \text{J} = \frac{1}{2}v^2$$

2. Evaluate v $v = 0.104$ mm/s

89* •• A woman whose weight on earth is 500 N is lifted to a height two earth radii above the surface of the earth. Her weight will (a) decrease to one-half of the original amount. (b) decrease to one-quarter of the original amount. (c) decrease to one-third of the original amount. (d) decrease to one-ninth of the original amount.

(d) g depends on $1/r^2$.

93* •   Uranus has a moon, Umbriel, whose mean orbital radius is 2.67×10^8 m and whose period is 3.58×10^5 s. (*a*) Find the period of another of Uranus's moons, Oberon, whose mean orbital radius is 5.86×10^8 m. (*b*) Use the known value of G to find the mass of Uranus.

(*a*) Use Equ. 11-2; $T_O = T_U(R_O/R_U)^{3/2}$

$T_O = (3.58 \times 10^5 \text{ s})(5.86/2.67)^{3/2} = 1.16 \times 10^6$ s

(*b*) $M = 4\pi^2 R^3/GT^2$

$M_U = 4\pi^2(2.67 \times 10^8)^3/G(3.58 \times 10^5)^2$ kg

$= 8.79 \times 10^{25}$ kg

97* ••   A uniform sphere of radius 100 m and density 2000 kg/m^3 is in free space far from other massive objects. (*a*) Find the gravitational field outside of the sphere as a function of r. (*b*) Find the gravitational field inside the sphere as a function of r.

(*a*) $g = -GM/r^2$; $M = 4\pi\rho R^3/3$

$M = 8.38 \times 10^9$ kg; $g = -0.559/r^2$

(*b*) Use Equ. 11-27

$g = -5.59 \times 10^{-7} r$

101* ••   A satellite is circling around the moon (radius 1700 km) close to the surface at a speed v. A projectile is launched from the moon vertically up at the same initial speed v. How high will it rise?

1. From Problem 55, $v^2 = v_e^2/2$; $\tfrac{1}{2}mv_e^2 = GmM_m/R_m$

$\tfrac{1}{2}v^2 = \tfrac{1}{2}GM_m/R_m = GM_m[1/R_m - 1/(R_m + h)]$

2. Solve for h

$h = R_m = 1700$ m

105* ••   A hole is drilled from the surface of the earth to its center as in Figure 11-27. Ignore the earth's rotation and air resistance. (*a*) How much work is required to lift a particle of mass m from the center of the earth to the earth's surface? (*b*) If the particle is dropped from rest at the surface of the earth, what is its speed when it reaches the center of the earth? (*c*) What is the escape speed for a particle projected from the center of the earth? Express your answers in terms of m, g, and R_E.

(*a*) From Equ. 11-27, $F = GmM_E r/R_E^3 = gmr/R_E$

$$W = \int_0^{R_E} F\, dr = \frac{gm}{R_E}\int_0^{R_E} r\, dr = \frac{gmR_E}{2}$$

(*b*) Use energy conservation; $\tfrac{1}{2}mv^2 = W$

$v = \sqrt{gR_E}$

(*c*) $E_{esc} = W + \tfrac{1}{2}mv_e^2 = \tfrac{1}{2}mv_{esc}^2$

$v_{esc}^2 = gR_E + 2gR_E$; $v_{esc} = (3gR_E)^{1/2} = 13.7$ km/s

---

109* •••   A uniform sphere of mass M is located near a thin, uniform rod of mass m and length L as in Figure 11-28. Find the gravitational force of attraction exerted by the sphere on the rod. (see Problem 72)

We shall determine the force exerted by the rod on the sphere and then use Newton's third law. The sphere is equivalent to a point mass m located at the sphere's center.

1. See Example 11-8; set $x_0 = a + L/2$

$F = GmM/[(a+L/2)^2 - L^2/4] = GmM/[a(a+L)]$

2. By Newton's third law, this is the force on the rod

CHAPTER 12

Static Equilibrium and Elasticity

1* • True or false: (*a*) $\Sigma F = 0$ is sufficient for static equilibrium to exist. (*b*) $\Sigma F = 0$ is necessary for static equilibrium to exist. (*c*) In static equilibrium, the net torque about any point is zero. (*d*) An object is in equilibrium only when there are no forces acting on it.

(*a*) False (*b*) True (*c*) True (*d*) False

5* • Misako wishes to measure the strength of her biceps muscle by exerting a force on a test strap as shown in Figure 12-25. The strap is 28 cm from the pivot point at the elbow, and her biceps muscle is attached at a point 5 cm from the pivot point. If the scale reads 18 N when she exerts her maximum force, what force is exerted by the biceps muscle?

Apply $\Sigma \tau = 0$ about the pivot $28 \times 18 = 5 \times F; F = 101$ N

9* • If the acceleration of gravity is not constant over an object, is it the center of mass or the center of gravity that is the pivot point when the object is balanced?

The center of gravity is then the pivot point for balance.

13* •• A square plate is produced by welding together four smaller square plates, each of side *a* as shown in Figure 12-28. Plate 1 weighs 40 N; plate 2, 60 N; plate 3, 30 N; and plate 4, 50 N. Find the center of gravity (x_{cg}, y_{cg}).

1. Use Equ. 12-3 to find x_{cg} $(1/2)a \times 100 + (3/2)a \times 80 = 180 \times x_{cg}; x_{cg} = 0.944a$

2. Similarly, find y_{cg} $(1/2)a \times 90 + (3/2)a \times 90 = 180 \times y_{cg}; y_{cg} = a$

17* • Figure 12-31 shows a 25-foot sloop. The mast is a uniform pole of 120 kg and is supported on the deck and held fore and aft by wires as shown. The tension in the forestay (wire leading to the bow) is 1000 N. Determine the tension in the backstay and the force that the deck exerts on the mast. Is there a tendency for the mast to slide forward or aft? If so, where should a block be placed to prevent the mast from moving?

1. Find θ_F, the angle of the forestay with vertical $\theta_F = \tan^{-1}(2.74/4.88) = 29.3°$

2. Apply $\Sigma \tau = 0$ with pivot at bottom of the mast $2.74 \times 1000 \times \cos 29.3° = 4.88 T_B \cos 45°; T_B = 692$ N

3. Apply $\Sigma F = 0$ to mast; F_D = force exerted by the deck on the mast $692 \times \sin 45° - 1000 \times \sin 29.3° = F_{Dx}; F_{Dx} = 0$

$F_{Dy} = 692 \times \cos 45° + 1000 \times \cos 29.3° + 120g = 2539$ N

21* •• A 3-m board of mass 5 kg is hinged at one end. A force F is applied vertically at the other end to lift a 60-kg block, which rests on the board 80 cm from the hinge, as shown in Figure 12-35. (*a*) Find the magnitude of the force needed to hold the board stationary at $\theta = 30°$. (*b*) Find the force exerted by the hinge at this angle. (*c*) Find the magnitude of the force F and the force exerted by the hinge if F is exerted perpendicular to the board when $\theta = 30°$.

(*a*) Apply $\Sigma \tau = 0$ about the hinge; cos 30° factors cancel

$3F = (0.8 \times 60 + 1.5 \times 5)9.81$ N·m; $F = 181.5$ N

(*b*) Use $\Sigma F = 0$

$F_H + 181.5 - 65 \times 9.81 = 0$; $F_H = 456$ N

(*c*) Apply $\Sigma \tau = 0$

$3F = (48 + 7.5) \times 9.81 \times \cos 30°$; $F = 157.2$ N

Use $\Sigma F_x = 0$, $\Sigma F_y = 0$

$F_{Hy} = 65 \times 9.81 - 157.2 \cos 30° = 501.5$ N;

$F_{Hx} = 157.2 \sin 30° = 78.6$ N; $F_H = (78.6\, i + 501.5\, j)$ N

25* •• As a farewell prank on their alma mater, Sharika and Chico decide to liberate thousands of marbles in the hallway during final exams. They place a 2-m × 1-m × 1-m box on a hinged board, as in Figure 12-38, and fill it with marbles. When the building is perfectly silent, they slowly lift one end of the plank, increasing θ, the angle of the incline. If the coefficient of static friction is large enough to prevent the box from slipping, at what angle will the box tip? (Assume that the marbles stay in the box until it tips over.)

The box will tip when its center of mass is no longer above the base of the box. So $\theta = \tan^{-1}(0.5) = 26.6°$.

29* •• The diving board shown in Figure 12-40 has a mass of 30 kg. Find the force on the supports when a 70-kg diver stands at the end of the diving board. Give the direction of each support force as a tension or a compression.

1. Use $\Sigma \tau = 0$ about the end support as a pivot to find the force on the middle support

$(4.2 \times 70 + 2.1 \times 30)g = 1.2F$; $F = 2920$ N, compression

2. $\Sigma \tau = 0$ about the right support to find F of the end support

$1.2F_{end} = (0.9 \times 30 + 3 \times 70)g$; $F_{end} = 1940$ N, tension

---

33* •• A cylinder of mass M and radius R rolls against a step of height h as shown in Figure 12-42. When a horizontal force F is applied to the top of the cylinder, the cylinder remains at rest. (*a*) What is the normal force exerted by the floor on the cylinder? (*b*) What is the horizontal force exerted by the corner of the step on the cylinder? (*c*) What is the vertical component of the force exerted by the corner on the cylinder?

(*a*) Take moments about the step's corner; see Example 12-4 for the moment arm of Mg and F_n

$F(2R - h) = (Mg - F_n)(2Rh - h^2)^{1/2}$; solve for F_n

$F_n = Mg - F[(2R - h)/h]^{1/2}$

(*b*) Use $\Sigma F_x = 0$

$F_{x,corn} = -F$

(*c*) Use $\Sigma F_y = 0$

$F_n - Mg + F_{y,corn} = 0$; $F_{y,corn} = F[(2R - h)/h]^{1/2}$

37* ••• A uniform log with a mass of 100 kg, a length of 4 m, and a radius of 12 cm is held in an inclined position, as shown in Figure 12-45. The coefficient of static friction between the log and the horizontal surface is 0.6. The log is on the verge of slipping to the right. Find the tension in the support wire and the angle the wire makes with the vertical wall.

We shall use the following nomenclature: T = the tension in the wire; F_n = the normal force of the surface; $f_s = \mu_s F_n$ = the force of static friction. We shall also use $\Sigma\tau = 0$ and use the point where the wire is attached to the log as the pivot. Taking this as the origin, the center of mass of the log is at the coordinates

$(-2\cos 20° + 0.12\sin 20°, -2\sin 20° - 0.12\cos 20°) = (-1.838, -0.797)$.

The point of contact with the floor is at $(-3.676, -1.594)$.

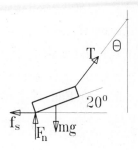

1. Use $\Sigma F = 0$; see the free body diagram	$T\sin\theta - \mu_s F_n = 0$;
	$T\cos\theta + F_n - mg = 0$
2. Apply $\Sigma\tau = 0$ about the origin	$1.838mg - 3.676F_n - 1.142\mu_s F_n = 0$
3. Solve for F_n; $m = 100$ kg, $\mu_s = 0.6$	$F_n = 389$ N
4. Insert 225 N $= F_n$ into the force equations	$T\sin\theta = 233$ N; $T\cos\theta = 592$ N;
	$\theta = \tan^{-1}(0.394) = 21.5°$
5. Evaluate T	$T = (233/\sin 21.5°)$ N $= 636$ N

41* •   A uniform cube of side a and mass M rests on a horizontal surface. A horizontal force F is applied to the top of the cube as in Figure 12-48. This force is not sufficient to move or tip the cube. (*a*) Show that the force of static friction exerted by the surface and the applied force constitute a couple, and find the torque exerted by the couple. (*b*) This couple is balanced by the couple consisting of the normal force exerted by the surface and the weight of the cube. Use this fact to find the effective point of application of the normal force when $F = Mg/3$. (*c*) What is the greatest magnitude of F for which the cube will not tip?

(*a*) The cube is stationary. Therefore $f_s = -F$, and the torque of that couple is Fa.

(*b*) Let x = the distance from the point of application of F_n to the center of the cube. Now, $F_n = Mg$, so $Mgx = Fa$; $x = Fa/Mg$. If $F = Mg/3$, then $x = a/3$.

(*c*) Note that $x_{max} = a/2$. The cube will tip if $F > Mg/2$.

45* ••   A massless ladder of length L leans against a smooth wall making an angle θ with the horizontal floor. The coefficient of friction between the ladder and the floor is μ_s. A man of mass M climbs the ladder. What height h can he reach before the ladder slips?

The ladder and the forces acting on it are shown in the diagram. Since the wall is smooth, the force the wall exerts on the ladder must be horizontal.

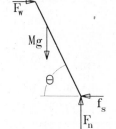

1. Use $\Sigma F = 0$	$F_n = Mg$; $f_s = F_w$
2. Apply $\Sigma\tau = 0$ about the bottom of the ladder	$F_w L\sin\theta = Mgx\cos\theta$
3. Solve for x; use $f_s = \mu_s F_n = \mu_s Mg = F_w$	$x = \mu_s L\tan\theta$; $h = x\sin\theta$
	$= \mu_s L\tan\theta\sin\theta$

49* ••   A ladder rests against a frictionless vertical wall. The coefficient of static friction between ladder and the floor is 0.3. What is the smallest angle at which the ladder will remain stationary?

Using the notation of Problem 45, we have $F_n = mg$ and $f_s = \mu_s mg = F_w$. Now take moments about the bottom of the ladder. This gives $\frac{1}{2}Lmg\cos\theta = Lf_w\sin\theta = Lmg\mu_s\sin\theta$. Solve for θ; $\theta = \tan^{-1}(1/2\mu_s) = 59°$.

53* •   A 50-kg ball is suspended from a steel wire of length 5 m and radius 2 mm. By how much does the wire stretch?

1. Find the stress

$F = mg = 490.5$ N; $A = \pi r^2 = 1.26 \times 10^{-5}$ m^2;

$F/A = 3.9 \times 10^7$ N/m^2

2. From Equ. 12-7 and Table 12-1 find ΔL

$\Delta L = (5 \times 3.9 \times 10^7/2 \times 10^{11})$ m $= 0.976$ mm

---

57* ••   A steel wire of length 1.5 m and diameter 1 mm is joined to an aluminum wire of identical dimensions to make a composite wire of length 3.0 m. What is the length of the composite wire if it is used to support a mass of 5 kg?

1. Find the stress in each wire

$F/A = (5 \times 9.81 \times 4/\pi \times 10^{-6})$ N/m$^2 = 6.245 \times 10^7$ N/m^2

2. Find $\Delta L = \Delta L_S + \Delta L_A$

$\Delta L_S = 1.5 \times 6.245 \times 10^7/2 \times 10^{11}$ m $= 0.468$ mm

$\Delta L_A = 0.468(200/70)$ mm $= 1.338$ mm; $\Delta L = 1.81$ mm

$L = 3.0$ m $+ \Delta L = 3.0018$ m

61* ••   A building is to be demolished by a 400-kg steel ball swinging on the end of a 30-m steel wire of diameter 5 cm hanging from a tall crane. As the ball is swung through an arc from side to side, the wire makes an angle of 50° with the vertical at the top of the swing. Find the amount by which the wire is stretched at the bottom of the swing.

1. Find the tension at bottom of the swing

$F = mg + mv^2/R$; $v^2/R = 2g(1 - \cos\theta) = 0.714g$

$F = 1.714mg = 6727$ N

2. $\Delta L = FL/AY$; $A = \pi \times 25 \times 10^{-4}/4$ m^2 $\Delta L = (6727 \times 30/1.96 \times 10^{-3} \times 2 \times 10^{11})$ m $= 0.515$ mm

   $= 1.96 \times 10^{-3}$ m^2

65* •••   It is apparent from Table 12-2 that the tensile strength of most materials is two to three orders of magnitude smaller than Young's modulus. Consequently, these materials, e.g., aluminum, will break before the strain exceeds even 1%. For nylon, however, the tensile strength and Young's modulus are approximately equal. If a nylon line of unstretched length L_0 and cross section A_0 is subjected to a tension T, the cross section may be substantially less than A_0 before the line breaks. Under these conditions, the tensile stress T/A may be significantly greater than T/A_0. Derive an expression that relates the area A to the tension T, A_0, and Young's modulus Y.

Assume constant volume of the line. Then $LA = $ constant, and taking differentials $L\Delta A + A\Delta L = 0$ or $\Delta L/L = -\Delta A/A$. But $\Delta L/L = T/AY$, so $\Delta A = -T/Y$. Thus $A = A_0 + \Delta A = A_0 - T/Y$.

69* ••   Sit in a chair with your back straight. Now try to stand up without leaning forward. Explain why you cannot do it.

The body's center of gravity must be above the feet.

73* •   A block and tackle is used to support a mass of 120 kg as shown in Figure 12-56. (*a*) What is the tension in the rope? (*b*) What is the mechanical advantage of this device?

(*a*) With this arrangement, the mass is supported by three ropes. Thus $T = 120g/3 = 392$ N.

(*b*) The mechanical advantage is 3.

77* ••   A uniform box of mass 8 kg that is twice as tall as it is wide rests on the floor of a truck. What is the maximum coefficient of static friction between the box and floor so that the box will slide toward the rear of the truck rather than tip when the truck accelerates on a level road?

Proceed as in the previous problem. The box will tip if $\mu_s > 0.5$, so it must have $\mu_s < 0.5$.

81* ••   A cube of mass M leans against a frictionless wall making an angle θ with the floor as shown in Figure 12-61. Find the minimum coefficient of static friction μ_s between the cube and the floor that allows the cube to stay at rest.

The figure alongside shows the location of the cube's center of mass and the forces acting on the cube. The moment arm of the couple formed by the normal force, F_n, and Mg is $d = (a/\sqrt{2})\sin(45° - \theta)$. The opposing couple is formed by the friction force f_s and the force exerted by the wall.

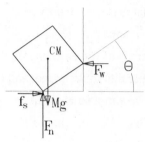

1. Set $\Sigma \tau = 0$ $(Mga/\sqrt{2})\sin(45° - \theta) = f_s a \sin \theta = \mu_s Mga \sin \theta$
2. Solve for μ_s; $\sin(\alpha + \beta) = \sin \alpha \cos \beta + \sin \beta \cos \alpha$ $\mu_s = \frac{1}{2}(\cot \theta - 1)$

85* ••   A uniform cube can be moved along a horizontal plane either by pushing the cube so that it slips or by turning it over ("rolling"). What coefficient of kinetic friction μ_k between the cube and the floor makes both ways equal in terms of the work needed?

To "roll" the cube one must raise its center of mass from $y = a/2$ to $y = \sqrt{2}a/2$, where a is the cube length. Thus the work done is $W = \frac{1}{2}mga(\sqrt{2} - 1) = 0.207mga$. Since no work is done as the cube flops down, this is the work done to move the cube a distance a. Now set $0.207mga = Fa = \mu_k mga$ = work done against friction in moving a distance a. Thus $\mu_k = 0.207$.

89* ••   A thin rod 60 cm long is balanced 20 cm from one end when a mass of $2m + 2$ g is at the end nearest the pivot and a mass of m at the opposite end (Figure 12-66a). Balance is again achieved if the mass $2m + 2$ g is replaced by the mass m and no mass is placed at the other end (Figure 12-66b). Determine the mass of the rod.

1. Take moments about pivot for initial condition $20(m + 2) = 40m + 10M$
2. Take moments about pivot for second condition $20m = 10M; M = 2m$
3. Solve for m and M using first equation $m = 1$ g; $M = 2$ g

---

93* ••   If a train travels around a bend in the railbed too fast, the freight cars will tip over. Assume that the cargo portions of the freight cars are regular parallelepipeds of uniform density and 1.5×10^4 kg mass, 10 m long, 3.0 m high, and 2.20 m wide, and that their base is 0.65 m above the rails. The axles are 7.6 m apart, each 1.2 m from the ends of the boxcar. The separation between the rails is 1.55 m. Find the maximum safe speed of the train if the radius of curvature of the bend is (a) 150 m, and (b) 240 m.

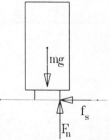

The box car and rail are shown in the drawing. At the critical speed, the normal force is entirely on the outside rail. As indicated, the center of gravity is 0.775 m from that rail and 2.15 m above it. To find the speed at which this situation prevails, we take moments about the center of gravity.

Set $\Sigma\tau = 0$ about the car's center of gravity $0.775mg = 2.15mv^2/R$

(a) Solve for v with $R = 150$ m $v = (0.775 \times 150 \times 9.81/2.15)^{\frac{1}{2}}$ m/s = 23 m/s = 83 km/h

(b) Find v for $R = 240$ m $v = 29.1$ m/s = 105 km/h

97* ••• A uniform sphere of radius R and mass M is held at rest on an inclined plane of angle θ by a horizontal string, as shown in Figure 12-68. Let $R = 20$ cm, $M = 3$ kg, and $\theta = 30°$. (a) Find the tension in the string. (b) What is the normal force exerted on the sphere by the inclined plane? (c) What is the frictional force acting on the sphere?

There are four forces acting on the sphere: its weight, mg; the normal force of the plane, F_n; the frictional force, f, acting parallel to the plane; and the tension in the string, T.

(a) 1. Take moments about the center of the sphere $TR = fR; \ T = f$

   2. Set $\Sigma F_x = 0$, where x is along the plane $T \cos \theta + f = mg \sin \theta; \ T = mg \sin \theta/(1 + \cos \theta)$

   3. Evaluate T $T = 3 \times 9.81 \times 0.5/(1 + 0.866) = 7.89$ N

(b) Set $\Sigma F_y = 0$ $F_n = mg \cos \theta + T \sin \theta = 29.4$ N

(c) $f = T$ $f = 7.89$ N

101*••• A solid cube of side length a balanced atop a cylinder of diameter d is in unstable equilibrium if $d \ll a$ and is in stable equilibrium if $d \gg a$ (Figure 12-71). Determine the minimum value of d/a for which the cube is in stable equilibrium.

Consider a small rotational displacement, $\delta\theta$, of the cube of the figure below from equilibrium. This shifts the point of contact between cube and cylinder by $R\delta\theta$, where $R = d/2$. As a result of that motion, the cube itself is rotated through the same angle $\delta\theta$, and so its center is shifted in the same direction by the amount $(a/2)\delta\theta$, neglecting higher order terms in $\delta\theta$. If the displacement of the cube's center of mass is less than that of the point of contact, the torque about the point of contact is a restoring torque, and the cube will return to its equilibrium position. If, on the other hand, $(a/2)\delta\theta > (d/2)\delta\theta$, then the torque about the point of contact due to mg is in the direction of $\delta\theta$, and will cause the displacement from equilibrium to increase. We see that the minimum value of d/a for stable equilibrium is $d/a = 1$.

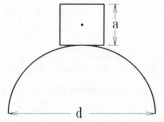   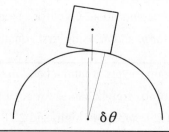

CHAPTER **13**

Fluids

1* • A copper cylinder is 6 cm long and has a radius of 2 cm. Find its mass.

Find the volume, then $m = \rho V$ $\qquad V = \pi r^2 h = 75.4 \times 10^{-6}$ m³; $m = 8.93 \times 10^3 V = 0.673$ kg

---

5* •• A 60-mL flask is filled with mercury at 0°C (Figure 13-22). When the temperature rises to 80°C, 1.47 g of mercury spills out of the flask. Assuming that the volume of the flask is constant, find the density of mercury at 80°C if its density at 0°C is 13,645 kg/m³.

1. Write ρ' in terms of ρ_0, V, and Δm $\qquad m = \rho_0 V$; $m' = m - \Delta m = \rho' V$; $\rho' = \rho_0 - \Delta m/V$

2. Evaluate ρ' $\qquad\qquad\qquad\qquad\qquad \rho' = 13{,}621$ kg/m³

9* • Find (a) the absolute pressure and (b) the gauge pressure at the bottom of a swimming pool of depth 5.0 m.

(a) $P = P_{at} + \rho g h$ $\qquad\qquad P = (1.01 \times 10^5 + 9.81 \times 10^3 \times 5) = 1.5 \times 10^5$ N/m²

$\qquad\qquad\qquad\qquad\qquad\qquad\qquad\quad = 1.5$ atm

(b) $P_{gauge} = P - P_{at}$ $\qquad\qquad P_{gauge} = 0.5$ atm

---

13* • What pressure is required to reduce the volume of 1 kg of water from 1.00 L to 0.99 L?

Use Equ. 13-6 $\qquad\qquad\qquad\qquad \Delta P = 2.0 \times 10^9 \times 10^{-2} = 2 \times 10^7$ Pa ≈ 200 atm

---

17* •• Many people have imagined that if they were to float the top of a flexible snorkel tube out of the water, they would be able to breathe through it while walking underwater (Figure 13-24). However, they generally do not reckon with just how much water pressure opposes the expansion of the chest and the inflation of the lungs. Suppose you can just breathe while lying on the floor with a 400-N weight on your chest. How far below the surface of the water could your chest be for you still to be able to breathe, assuming your chest has a frontal area of 0.09 m²?

$P = \rho_w g h = F/A$; $h = F/\rho_w g A$ $\qquad\qquad h = (400/1 \times 10^3 \times 9.81 \times 0.09)$ m $= 0.453$ m $= 45.3$ cm

---

21* ••• The volume of a cone of height h and base radius r is $V = \pi r^2 h/3$. A conical vessel of height 25 cm resting on its base of radius 15 cm is filled with water. (a) Find the volume and weight of the water in the vessel. (b) Find the force exerted by the water on the base of the vessel. Explain how this force can be greater than the weight of the water.

(a) $w = \rho g V$; V is given $V = 5.89 \times 10^{-3}$ m^3; $w = 5.89 \times 9.81 = 57.8$ N

(b) $F = PA = \rho g h A$ $F = \rho g h \pi r^2 = 3 \rho g V = 3 \times 57.8$ N $= 173$ N

This occurs in the same way that the force on Pascal's barrel >> the weight of water in the tube. The downward force on the base is also the result of the downward component of the force exerted by the slanting walls of the cone on the water.

25* ••   A fishbowl rests on a scale. The fish suddenly swims upward to get food. What happens to the scale reading? Nothing. The fish is in neutral buoyancy, so the upward acceleration of the fish is balanced by the downward acceleration of the displaced water.

29* •   A 500-g piece of copper (specific gravity 9.0) is suspended from a spring scale and is submerged in water (Figure 13-26). What force does the spring scale read?

$w = \rho_{Cu} V g$; $w' = \rho_{Cu} V g - \rho_w V g = (\rho_{Cu} - \rho_w) w / \rho_{Cu}$ $w' = 0.5 \times 9.81(7.93/8.93)$ N $= 4.36$ N

33* ••   An object floats on water with 80% of its volume below the surface. The same object when placed in another liquid floats on that liquid with 72% of its volume below the surface. Determine the density of the object and the specific gravity of the liquid.

$\rho = \rho_w(V'/V) = \rho_L(V''/V)$ (see Example 13-7) $\rho = 0.8\rho_w = 800$ kg/m^3;

$\rho_L/\rho_w = 0.8/0.72 = 1.11 =$ sp. gravity

37* ••   A helium balloon lifts a basket and cargo of total weight 2000 N under standard conditions, in which the density of air is 1.29 kg/m^3 and the density of helium is 0.178 kg/m^3. What is the minimum volume of the balloon?

1. Set $F_B =$ total weight $= \rho_{He} g V + 2000$ N $1.29 g V = 0.178 g V + 2000$ N

2. Solve for and evaluate V $V = [2000/9.81(1.29 - 0.178)]$ m$^3 = 183$ m^3

41* •••   A ship sails from seawater (specific gravity 1.025) into fresh water and therefore sinks slightly. When its load of 600,000 kg is removed, it returns to its original level. Assuming that the sides of the ship are vertical at the water line, find the mass of the ship before it was unloaded.

Let $V =$ displacement of ship in the two cases, m be the mass of ship without load, Δm be the load.

1. Write condition for floating in both cases $(m + \Delta m) = \rho_{sw} V$; $m = \rho_w V$

2. Solve for and evaluate $m + \Delta m$ $m + \Delta m = \Delta m \rho_{sw}/(\rho_{sw} - \rho_w) = 2.06 \times 10^7$ kg

---

45* ••   When water emerges from a faucet, the stream narrows as the water falls. Explain why.

Pressure within the stream diminishes as the velocity of the stream increases.

49* ••   Blood flows in an aorta of radius 9 mm at 30 cm/s. (a) Calculate the volume flow rate in liters per minute. (b) Although the cross-sectional area of a capillary is much smaller than that of the aorta, there are many capillaries, so their total cross-sectional area is much larger. If all the blood from the aorta flows into the capillaries and the speed of flow through the capillaries is 1.0 mm/s, calculate the total cross-sectional area of the capillaries.

(a) $I_V = A v$ $I_V = (\pi \times 81 \times 10^{-6} \times 0.3)$ m^3/s $= 7.63 \times 10^{-5}$ m^3/s

                                                      $= 4.58$ L/min

(b) $A_{cap}v_{cap} = I_v$; $A_{cap} = I_v/v_{cap}$ $\qquad\qquad\qquad\qquad$ $A_{cap} = 7.63 \times 10^{-5}/10^{-3}$ m^2 = 7.63×10^{-2} m^2 = 763 cm^2

53* •• Water flows through a Venturi meter like that in Example 13-9 with a pipe diameter of 9.5 cm and a constriction diameter of 5.6 cm. The U-tube manometer is partially filled with mercury. Find the flow rate of the water in the pipe of 9.5-cm diameter if the difference in the mercury level in the U-tube is 2.40 cm.

$$v_1 = \sqrt{\frac{2\rho_L gh}{\rho_F(r-1)}} \; ; \; r = \frac{R_1^2}{R_2^2} \text{ (see Example 13-9)} \qquad v_1 = \sqrt{\frac{2 \times 13.6 \times 10^3 \times 9.81 \times 2.4 \times 10^{-2}}{10^3(2.88-1)}} \text{ m/s}$$

$$v_1 = 1.85 \text{ m/s}$$

$$I_v = \pi r^2 v_1 \qquad\qquad\qquad\qquad I_v = (\pi \times 0.095^2 \times 1.85/4) \text{ m}^3\text{/s} = 13.1 \text{ L/s}$$

57* • A horizontal tube with an inside diameter of 1.2 mm and a length of 25 cm has water flowing through it at 0.30 mL/s. Find the pressure difference required to drive this flow if the viscosity of water is 1.00 mPa·s.

Use Equ. 13-23 $\qquad\qquad\qquad\qquad\qquad\qquad \Delta P = (8 \times 10^{-3} \times 0.25 \times 0.3 \times 10^{-6}/\pi \times 0.6^4 \times 10^{-12})$ Pa

$\qquad\qquad\qquad\qquad\qquad\qquad\qquad\qquad\qquad\qquad = 1.47$ kPa

---

61* • A glass of water has ice cubes floating in it. What happens to the water level when the ice melts?

The water level remains constant.

65* •• In Example 13-9, the fluid is accelerated to a greater speed as it enters the narrow part of the pipe. Identify the forces that act on the fluid to produce this acceleration?

The force acting on the fluid is the difference in pressure between the wide and narrow parts times the area of the narrow part.

69* • The top of a card table is 80 cm × 80 cm. What is the force exerted on it by the atmosphere? Why doesn't the table collapse?

The net force is zero. Neglecting the thickness of the table, the atmospheric pressure is the same above and below the surface of the table.

73* • A solid cubical block of side length 0.6 m is suspended from a spring balance. When the block is in water, the spring balance reads 80% of the reading when the block is in air. Determine the density of the block.

Use Equ. 13-12; $\rho = \rho_w(w_0/\Delta w)$ $\qquad\qquad\qquad \rho = 1000(1/0.2) = 5000$ kg/m^3

77* • Suppose that when you are floating in fresh water, 96% of your body is submerged. What is the volume of water your body displaces when it is fully submerged?

The mass of your body divided by the density of water; e.g., (60 kg)/(1000 kg/m^3) = 0.06 m^3.

81* •• A beaker filled with water is balanced on the left cup of a scale. A cube 4 cm on a side is attached to a string and lowered into the water so that it is completely submerged. The cube is not touching the bottom of the beaker. A weight m is added to the system to retain equilibrium. What is the weight m and on which cup of the balance is it added?

The additional weight on the beaker side equals the weight of the displaced water, i.e., 64 g. This is the mass that must be placed on the other cup to maintain balance.

85* •• Figure 13-33 is a sketch of an *aspirator*, a simple device that can be used to achieve a partial vacuum in a reservoir connected to the vertical tube at B. An aspirator attached to the end of a garden hose may be used to deliver soap or fertilizer from the reservoir. Suppose that the diameter at A is 2.0 cm and at C, where the water exits to the atmosphere, it is 1.0 cm. If the flow rate is 0.5 L/s and the gauge pressure at A is 0.187 atm, what diameter of the constriction at B will achieve a pressure of 0.1 atm in the container?

Since it is not given, we shall neglect the difference in height between the centers of the pipes at A and B.

1. Find v_A $v_A = I_v/A_A = 5 \times 10^{-4}/\pi \times 10^{-4}$ m/s $= 5/\pi$ m/s

2. Use Equ. 13-18 to find v_B^2 $v_B^2 = (1.087 \times 1.01 \times 10^5 + 500 \times 25/\pi^2)/500$ m^2/s^2

                                                          $= 222$ m^2/s^2

3. $v_B = 5 \times 10^{-4}/\pi r_B^2 = 14.9$ m/s; solve for r_B $r_B = (5 \times 10^{-4}/14.9\pi)^{\frac{1}{2}}$ m $= 3.27$ mm; $d = 6.54$ mm

---

89* •• A rectangular dam 30 m wide supports a body of water to a height of 25 m. (*a*) Neglecting atmospheric pressure, find the total force due to water pressure acting on a thin strip of the dam of height *dy* located at a depth *y*. (*b*) Integrate your result in part (*a*) to find the total horizontal force exerted by the water on the dam. (*c*) Why is it reasonable to neglect atmospheric pressure?

This problem is identical to Example 13-2.

(*a*) $dF = \rho g y L\, dy$.

(*b*) $F = \frac{1}{2}\rho g L h^2 = 9.20 \times 10^7$ N.

(*c*) Atmospheric pressure is exerted on each side of the dam and it can, therefore, be neglected.

93* •• A helium balloon can just lift a load of 750 N. The skin of the balloon has a mass of 1.5 kg. (*a*) What is the volume of the balloon? (*b*) If the volume of the balloon is twice that found in part (*a*), what is the initial acceleration of the balloon when it carries a load of 900 N?

(*a*) Write the condition for neutral buoyancy $V\rho_{air}g = 750 + 1.5g + V\rho_{He}g$

      Solve for V with $\rho_{air} = 1.293$, $\rho_{He} = 0.1786$ $V = 70$ m^3

(*b*) $F_{net} = F_B - mg$; $mg = 900 + 1.5g$ $F_{net} = [140(1.293 - 0.1786)g - 915]$ N $= 616$ N

      $a = F_{net}/m$ $a = [616/(900/g + 1.5)]$ m/s^2 $= 6.61$ m/s^2

97* •• A submarine has a total mass of 2.4×10^6 kg, including crew and equipment. The vessel consists of two parts, the pressure hull, which has a volume of 2×10^3 m^3, and the diving tanks, which have a volume of 4×10^2 m^3. When the sub cruises on the surface, the diving tanks are filled with air; when cruising below the surface, seawater is admitted into the tanks. (*a*) What fraction of the submarine's volume is above the water surface when the tanks are filled with air? (*b*) How much water must be admitted into the tanks to give the submarine neutral buoyancy? Neglect the mass of air in the tanks and use 1.025 as the specific gravity of seawater.

(*a*) Apply the expression of Example 13-7 $V'/V = \rho'/\rho_{SW}$; $V'/V = 1/1.025 = 0.9756$

      The fraction above the surface $= 1 - V'/V$ fraction above the surface $= 2.44\%$

(*b*) Apply the condition for neutral buoyancy $2.4 \times 10^3 \times 1.025 \times 10^3 = 2.4 \times 10^6 + 1.025 \times 10^3 V_{SW}$

      Solve for V_{SW} $V_{SW} = 58.5$ m^3

CHAPTER **14**

Oscillations

1* • Deezo the Clown slept in again. As he roller-skates out the door at breakneck speed on his way to a lunchtime birthday party, his superelastic suspenders catch on a fence post, and he flies back and forth, oscillating with an amplitude A. What distance does he move in one period? What is his displacement over one period?

In one period, he moves a distance $4A$. Since he returns to his initial position, his displacement is zero.

5* • True or false: (*a*) In simple harmonic motion, the period is proportional to the square of the amplitude. (*b*) In simple harmonic motion, the frequency does not depend on the amplitude. (*c*) If the acceleration of a particle is proportional to the displacement and oppositely directed, the motion is simple harmonic.

(*a*) False (*b*) True (*c*) True

9* • A particle of mass m begins at rest from $x = +25$ cm and oscillates about its equilibrium position at $x = 0$ with a period of 1.5 s. Write equations for (*a*) the position x versus the time t, (*b*) the velocity v versus t, and (*c*) the acceleration a versus t.

(*a*) $x = A \cos[(2\pi/T)t + \delta]$ ⟶ $x = 25 \cos(4.19t)$ cm

(*b*) $v = -A\omega \sin(\omega t)$ ⟶ $v = -105 \sin(4.19T)$ cm/s

(*c*) $a = -\omega^2 x$ ⟶ $a = -439 \cos(4.19t)$ cm/s^2

13* •• The period of an oscillating particle is 8 s. At $t = 0$, the particle is at rest at $x = A = 10$ cm. (*a*) Sketch x as a function of t. (*b*) Find the distance traveled in the first second, the next second, the third second, and the fourth second after $t = 0$.

(*a*) A sketch of $x = 10 \cos(\pi t/4)$ cm is shown

(*b*) In each case, $\Delta x = 10[\cos(\pi t_f/4) - \cos(\pi t_i/4)]$ cm

For $t_f = 1$ s, $t_i = 0$, $\Delta x = 2.93$ cm

For $t_f = 2$ s, $t_i = 1$ s, $\Delta x = 7.07$ cm

For $t_f = 3$ s, $t_i = 2$ s, $\Delta x = 7.07$ cm

For $t_f = 4$ s, $t_i = 3$ s, $\Delta x = 2.93$ cm

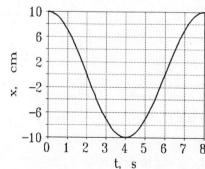

17* •   A particle moves in a circle of radius 40 cm with a constant speed of 80 cm/s. Find (*a*) the frequency of the motion, and (*b*) the period of the motion. (*c*) Write an equation for the *x* component of the position of the particle as a function of time *t*, assuming that the particle is on the positive *x* axis at time *t* = 0.

(*b*) $T = 2\pi r/v$ $T = \pi\,\mathrm{s} = 3.14\,\mathrm{s}$

(*a*) $f = 1/T$ $f = 1/\pi\,\mathrm{Hz} = 0.318\,\mathrm{Hz}$

(*c*) $x = 40\cos(2\pi ft + \delta)$ cm $x = 40\cos(2t + \delta)$ cm, where $\delta < \pi/2$

21* •   A 2.4-kg object is attached to a horizontal spring of force constant $k = 4.5$ kN/m. The spring is stretched 10 cm from equilibrium and released. Find its total energy.

$U = \frac{1}{2}kx^2$ $U = 4.5 \times 10^3 \times 0.1^2/2\,\mathrm{J} = 22.5\,\mathrm{J}$

25* •   An object oscillates on a spring with an amplitude of 4.5 cm. Its total energy is 1.4 J. What is the force constant of the spring?

$E = \frac{1}{2}kA^2$ $k = 2E/A^2 = 2.8/0.045^2\,\mathrm{N/m} = 1383\,\mathrm{N/m}$

29* •   A 2.4-kg object is attached to a horizontal spring of force constant $k = 4.5$ kN/m. The spring is stretched 10 cm from equilibrium and released. Find (*a*) the frequency of the motion, (*b*) the period, (*c*) the amplitude, (*d*) the maximum speed, and (*e*) the maximum acceleration. (*f*) When does the object first reach its equilibrium position? What is its acceleration at this time?

(*a*) $f = (1/2\pi)(k/m)^{\frac{1}{2}}$ $f = (1/2\pi)(4.5 \times 10^3/2.4)^{\frac{1}{2}}\,\mathrm{Hz} = 6.89\,\mathrm{Hz}$

(*b*) $T = 1/f$ $T = 0.145\,\mathrm{s}$

(*c*) A is given $A = 0.1\,\mathrm{m}$

(*d*) $v_{max} = A\omega = 2\pi Af$ $v_{max} = 4.33\,\mathrm{m/s}$

(*e*) $a_{max} = A\omega^2 = \omega v_{max}$ $a_{max} = 187\,\mathrm{m/s}^2$

(*f*) It reaches equilibrium after 1/4 period $t = T/4 = 36.25\,\mathrm{ms}$; at equilibrium, $a = 0$

33* •   A 4.5-kg object oscillates on a horizontal spring with an amplitude of 3.8 cm. Its maximum acceleration is 26 m/s². Find (*a*) the force constant *k*, (*b*) the frequency, and (*c*) the period of the motion.

(*a*) $k = ma_{max}/A$ $k = 4.5 \times 26/0.038\,\mathrm{N/m} = 3079\,\mathrm{N/m}$

(*b*) $\omega^2 = a_{max}/A; f = (a_{max}/A)^{\frac{1}{2}}/2\pi$ $f = (26/0.038)^{\frac{1}{2}}/2\pi\,\mathrm{Hz} = 4.16\,\mathrm{Hz}$

(*c*) $T = 1/f$ $T = 0.24\,\mathrm{s}$

37* ••   An object of unknown mass is hung on the end of an unstretched spring and is released from rest. If the object falls 3.42 cm before first coming to rest, find the period of the motion.

1. Take $E_i = 0$; $E_f = U_g + U_{spring}$ $-mg \times 0.0342 + \frac{1}{2}k \times 0.0342^2 = 0$

2. Solve for $k/m = \omega^2 = 4\pi^2/T^2$ $\omega^2 = 573.7\,\mathrm{rad}^2/\mathrm{s}^2$; $T = 0.262\,\mathrm{s}$

41* ••   In Problem 40, find the maximum amplitude of oscillation such that the stone will remain on the block. To remain on the block, the block's maximum downward acceleration must not exceed *g*.

1. Find a_{max} for the amplitude A $a_{max} = kA/m = 47.9A$

2. Set $a_{max} = g$ to determine A_{max} $A_{max} = 9.81/47.9\,\mathrm{m} = 20.5\,\mathrm{cm}$

45* •• A 1.5-kg object that stretches a spring 2.8 cm from its natural length when hanging at rest oscillates with an amplitude of 2.2 cm. (a) Find the total energy of the system. (b) Find the gravitational potential energy at maximum downward displacement. (c) Find the potential energy in the spring at maximum downward displacement. (d) What is the maximum kinetic energy of the object? (Choose $U = 0$ when the object is in equilibrium.)

Find k and then proceed as in the preceding problem. $k = mg/y_0 = 1.5 \times 9.81/0.028$ N/m $= 526$ N/m.

(a) $E = 263 \times 0.022^2$ J $= 0.127$ J. (b) $U_g = -1.5 \times 9.81 \times 0.022$ J $= -0.324$ J.

(c) $U_{spring} = (0.127 + 0.324)$ J $= 0.451$ J. (d) $K_{max} = \frac{1}{2}kA^2 = 263 \times 0.022^2$ J $= 0.127$ J.

49* •• The length of the string or wire supporting a pendulum increases slightly when its temperature is raised. How would this affect a clock operated by a simple pendulum?

The clock would run slow.

53* • A pendulum set up in the stairwell of a 10-story building consists of a heavy weight suspended on a 34.0-m wire. If $g = 9.81$ m/s^2, what is the period of oscillation?

Use Equ. 14-27 $T = 2\pi(34/9.81)^{1/2}$ s $= 11.7$ s

57* • A thin disk of mass 5 kg and radius 20 cm is suspended by a horizontal axis perpendicular to the disk through its rim. The disk is displaced slightly from equilibrium and released. Find the period of the subsequent simple harmonic motion.

1. Find I through the pivot; use parallel axis theorem $I = \frac{1}{2}MR^2 + MR^2 = 3MR^2/2 = 0.3$ kg·m^2

2. Apply Equ. 14-31 $T = 2\pi(I/MgD)^{1/2} = 2\pi(0.3/5 \times 9.81 \times 0.2)^{1/2}$ s $= 1.1$ s

61* •• Suppose the rod in Problem 60 has a mass of $2m$ (Figure 14-30). Determine the distance between the upper mass and the pivot point P such that the period of this physical pendulum is a minimum.

Folllow the same procedure as in Problem 60(a). Here $I_{cm} = mL^2/2 + 2mL^2/12 = 2mL^2/3$ and $I = 2mL^2/3 + 4mx^2$, where x is again the distance of the pivot from the center of the rod. The period is then $T = C[(2L^2/3 + 4x^2)/x]^{1/2}$, where C is a constant. Setting $dT/dx = 0$ gives $4x^2 = 2L^2/3$ and $x = L/\sqrt{6}$. The distance to the pivot from the nearer mass is then $d = L/2 - L/\sqrt{6} = 0.0918L$.

65* •• Figure 14-32 shows a uniform disk of radius $R = 0.8$ m and a 6-kg mass with a small hole a distance d from the disk's center that can serve as a pivot point. (a) What should be the distance d so that the period of this physical pendulum is 2.5 s? (b) What should be the distance d so that this physical pendulum will have the shortest possible period? What is this period?

(a) 1. Use Equ. 14-31; $T^2(mg/4\pi^2)d = I_{cm} + md^2$ $9.32d = \frac{1}{2} \times 6 \times 0.64 + 6d^2 = 1.92 + 6d^2$

   2. Solve the quadratic equation for d $d = 1.31$ m, $d = 0.244$ m; $d = 24.4$ cm

(b) Set $dT^2/dd = 0$ and solve for d $d^2 - \frac{1}{2}R^2 = 0$; $d = R/\sqrt{2} = 56.6$ cm

$$T = 2\pi\sqrt{\frac{I}{mgR/\sqrt{2}}};\qquad\qquad T = 2\pi\sqrt{\frac{R\sqrt{2}}{g}} = 2.1 \text{ s}$$

$$I = \frac{1}{2}mR^2 + m(\frac{1}{2}R^2) = mR^2$$

69* •• Two clocks have simple pendulums of identical lengths L. The pendulum of clock A swings through an arc of 10°; that of clock B swings through an arc of 5°. When the two clocks are compared one will find that (a) A runs slow compared to B. (b) A runs fast compared to B. (c) both clocks keep the same time. (d) the answer depends on the length L.

(a) The period of A is longer. (see Equ. 14-28)

73* • True or false: The energy of a damped, undriven oscillator decreases exponentially with time.

True

77* •• Show that the ratio of the amplitudes for two successive oscillations is constant for a damped oscillator.

From Equ. 14-36, $A(t)/A(t + T) = e^{-T/2\tau}$.

81* •• A damped mass–spring system oscillates at 200 Hz. The time constant of the system is 2.0 s. At $t = 0$, the amplitude of oscillation is 6.0 cm and the energy of the oscillating system is then 60 J. (a) What are the amplitudes of oscillation at $t = 2.0$ s and $t = 4.0$ s? (b) How much energy is dissipated in the first 2-s interval and in the second 2-s interval?

(a) $A(t) = A_0 e^{-t/2\tau}$

(b) $E(t) = E_0 e^{-t/\tau}$; $\Delta E = E_0(1 - e^{-t/\tau})$

$A(2) = 6e^{-0.5}$ cm $= 3.64$ cm; $A(4) = 6e^{-1}$ cm $= 2.21$ cm

$\Delta E_{0-2} = 60 \times 0.632$ J $= 37.9$ J;

$\Delta E_{2-4} = 37.9 \times 0.632$ J $= 24$ J

85* • Give some examples of common systems that can be considered to be driven oscillators.

The pendulum of a clock, a violin string when bowed, and the membrane of a loudspeaker can be considered driven oscillators.

89* •• A 2-kg object oscillates on a spring of force constant $k = 400$ N/m. The damping constant has a value of $b = 2.00$ kg/s. The system is driven by a sinusoidal force of maximum value 10 N and angular frequency $\omega = 10$ rad/s. (a) What is the amplitude of the oscillations? (b) If the driving frequency is varied, at what frequency will resonance occur? (c) What is the amplitude of oscillation at resonance? (d) What is the width of the resonance curve $\Delta\omega$?

(a) 1. Determine ω_0

2. Find A; use Equ. 14-49

(b) Resonance is at $\omega = \omega_0$

(c) Use Equ. 14-49 to find A_{res}

(d) From Equ. 14-39 and 14-45, $\Delta\omega = b/m$

$\omega_0 = (k/m)^{1/2} = 14.14$ rad/s

$A = 10/[4(200 - 100)^2 + 4 \times 100]^{1/2}$ m $= 0.05$ m $= 5.0$ cm

$\omega_{res} = 14.14$ rad/s

$A_{res} = 10/(4 \times 200)^{1/2} = 35.4$ cm

$\Delta\omega = 1$ rad/s

93* ••• Figure 14-37 shows a vibrating mass–spring system supported on a frictionless surface and a second equal mass that is moving toward the vibrating mass with velocity v. The motion of the vibrating mass is given by $x(t) = (0.1$ m$) \cos (40$ s$^{-1} t)$, where x is the displacement of the mass from its equilibrium position. The two masses collide elastically just as the vibrating mass passes through its equilibrium position traveling to the right. (a) What should be the velocity v of the second mass so that the mass–spring system is at rest following the elastic collision? (b) What is the velocity of the second mass after the elastic collision?

(a) 1. Apply conservation of energy and momentum

2. Since masses cancel we have

$Mv_{1i}^2 + Mv_{2i}^2 = Mv_{2f}^2$; $Mv_{1i} + Mv_{2i} = Mv_{2f}$

$(v_{2f} + v_{2i})(v_{2f} - v_{2i}) = v_{1i}^2$; $v_{2f} - v_{2i} = v_{1i}$

$v_{2f}^2 - v_{2i}^2 = v_{1i}^2$

3. Solve for v_{1i} $v_{1i} = 0$; the mass is at rest initially

(b) $v_{2f} = v_{1i}$; $v_{1i} = A\omega = 4$ m/s $v_{2f} = 4$ m/s

---

97* •• A lamp hanging from the ceiling of the club car in a train oscillates with period T_0 when the train is at rest. The period will be (match left and right columns)

1. greater than T_0 when   B. the train rounds a curve of radius R with speed v.

2. less than T_0 when   D. the train goes over a hill of radius of curvature R with constant speed.

3. equal to T_0 when   A. the train moves horizontally with constant velocity.

                                              C. the train climbs a hill of inclination θ at constant speed.

101* •• Pendulum A has a bob of of mass M_A and a length L_A; pendulum B has a bob of mass M_B and a length L_B. If the period of A is twice that of B, then (a) $L_A = 2L_B$ and $M_A = 2M_B$. (b) $L_A = 4L_B$ and $M_A = M_B$. (c) $L_A = 4L_B$ whatever the ratio M_A/M_B. (d) $L_A = \sqrt{2}L_B$ whatever the ratio M_A/M_B.

(c) T is independent of M; $L \propto T^2$.

105* •• A small particle of mass m slides without friction in a spherical bowl of radius r. (a) Show that the motion of the particle is the same as if it were attached to a string of length r. (b) Figure 14-38 shows a particle of mass m_1 that is displaced a small distance s_1 from the bottom of the bowl, where s_1 is much smaller than r. A second particle of mass m_2 is dislaced in the opposite direction a distance $s_2 = 3s_1$, where s_2 is also much smaller than r. If the particles are released at the same time, where do they meet? Explain.

(a) Since there is no friction, the only forces acting on the particle are mg and the normal force acting radially inward; the normal force is identical to the tension in a string of length r that keeps the particle moving in a circular path.

(b) The particles meet at the bottom. Since s_1 and s_2 are both much smaller than r, the particles behave like the bobs of simple pendulums of equal length and, therefore, have the same periods.

109* •• A wooden cube with edge a and mass m floats in water with one of its faces parallel to the water surface. The density of the water is ρ. Find the period of oscillation in the vertical direction if it is pushed down slightly.

1. Find the change in the buoyant force $dF_B = -\rho V g = -a^2 \rho g y$

2. Write the equation of motion $m(d^2y/dt^2) = -a^2\rho g y$; $d^2y/dt^2 = -(a^2\rho g/m)y$

3. Compare with Equs. 14-2 and 14-7 $\omega = a\sqrt{\rho g/m}$; $T = 2\pi/\omega = (2\pi/a)\sqrt{m/\rho g}$

113* •• An object of mass m_1 sliding on a frictionless horizontal surface is attached to a spring of force constant k and oscillates with an amplitude A. When the spring is at its greatest extension and the mass is instantaneously at rest, a second object of mass m_2 is placed on top of it. (a) What is the smallest value for the coefficient of static friction μ_s such that the second object does not slip on the first? (b) Explain how the total energy E, the amplitude A, the angular frequency ω, and the period T of the system are changed by placing m_2 on m_1, assuming that the friction is great enough so that there is no slippage.

(a) $\mu_{s,min} = a_{max}/g = \omega^2 A/g\mu_s = Ak/g(m_1 + m_2)$.

(b) A is unchanged; E is unchanged since $E = \frac{1}{2}kA^2$; ω is reduced by increasing the total mass; T is increased.

117* •• A small block of mass m_1 rests on a piston that is vibrating vertically with simple harmonic motion given by $y = A \sin \omega t$. (a) Show that the block will leave the piston if $\omega^2 A > g$. (b) If $\omega^2 A = 3g$ and $A = 15$ cm, at what time will the block leave the piston?

(a) At maximum upward extension, the block is momentarily at rest. Its downward acceleration is g. The downward acceleration of the piston is $\omega^2 A$. Therefore, if $\omega^2 A > g$, the block will separate from the piston.

(b) 1. $y = A \sin(\omega t)$; find a and critical ωt $a = -\omega^2 A \sin(\omega t) = -3g \sin(\omega t) = -g$; $\omega t = 0.34$ rad

     2. $\omega = (3g/0.15)^{\frac{1}{2}}$; find t $t = 0.34/14 = 0.0243$ s

---

121* ••• Repeat Problem 120 with $U(x) = U_0(\alpha^2 + 1/\alpha^2)$.

(a) A plot of $U(x)$ versus x/a is shown.

(b) $dU/dx = 0 = (2U_0/a)(\alpha - 1/\alpha^3)$; $\alpha_0 = 1$, $x_0 = a$.

(c) $U(x_0 + \varepsilon) = U_0[(1 + \beta)^2 + (1 + \beta)^{-2}]$; $\beta = \varepsilon/a$

(d) $U(x_0 + \varepsilon) \approx U_0(1 + 2\beta + \beta^2 + 1 - 2\beta + 3\beta^2) = \text{constant} + 4U_0\beta^2$;

     $U(x_0 + \varepsilon) = \text{constant} + 4U_0\varepsilon^2/a^2$

(e) For SHO, $U(\varepsilon) = \frac{1}{2}k\varepsilon^2$; so $k = 8U_0/a^2$; $f = (1/\pi a)(2U_0/m)^{\frac{1}{2}}$

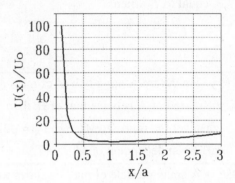

125* ••• A straight tunnel is dug through the earth as shown in Figure 14-46. Assume that the walls of the tunnel are frictionless. (a) The gravitational force exerted by the earth on a particle of mass m at a distance r from the center of the earth when $r < R_E$ is $F_r = -(GmM_E/R_E^3)r$, where M_E is the mass of the earth and R_E is its radius. Show that the net force on a particle of mass m at a distance x from the middle of the tunnel is given by $F_x = -(GmM_E/R_E^3)x$, and that the motion of the particle is therefore simple harmonic motion. (b) Show that the period of the motion is given by $T = 2\pi\sqrt{R_E/g}$ and find its value in minutes. (This is the same period as that of a satellite orbiting near the surface of the earth and is independent of the length of the tunnel.)

(a) From Figure 14-46, $F_x = F_r \sin \theta$; $\sin \theta = x/R$ $F_x = -(GmM_E/R_E^3)x$

(b) 1. Here $k_{eff} = (GmM_E/R_E^3)$; $T = 2\pi(m/k_{eff})^{\frac{1}{2}}$ $T = 2\pi(R_E^3/GM_E)^{\frac{1}{2}} = 2\pi(R_E/g)$

     2. Substitute appropriate numerical values $T = 2\pi(6.37 \times 10^6/9.81)^{\frac{1}{2}}$ s $= 5.06 \times 10^3$ s $= 84.4$ min

---

129* ••• In this problem, you will derive the expression for the average power delivered by a driving force to a driven oscillator (Figure 14-25). (a) Show that the instantaneous power input of the driving force is given by $P = Fv = -A\omega F_0 \cos \omega t \sin(\omega t - \delta)$. (b) Use the trigonometric identity $\sin(\theta_1 - \theta_2) = \sin \theta_1 \cos \theta_2 - \cos \theta_1 \sin \theta_2$ to show that the equation in (a) can be written $P = A\omega F_0 \sin \delta \cos^2 \omega t - A\omega F_0 \cos \delta \cos \omega t \sin \omega t$. (c) Show that the average value of the second term in your result for (b) over one or more periods is zero and that therefore $P_{av} = \frac{1}{2}A\omega F_0 \sin \delta$. (d) From Equation 14-50 for tan δ, construct a right triangle in which the side opposite the angle δ is $b\omega$ and the side adjacent is $m(\omega_0^2 - \omega^2)$, and use this triangle to show that

$$\sin \delta = \frac{b\omega}{\sqrt{m^2(\omega_0^2 - \omega^2)^2 + b^2\omega^2}} = \frac{b\omega A}{F_0} \, .$$

(e) Use your result for (d) to eliminate ωA so that the average power input can be written

$$P_{av} = \frac{1}{2}\frac{F_0^2}{b}\sin^2\delta = \frac{1}{2}\left[\frac{b^2\omega^2 F_0^2}{m^2(\omega_0^2 - \omega^2)^2 + b^2\omega^2}\right] \qquad (14\text{-}51).$$

(a) $F = F_0\cos\omega t$; $x(t) = A\cos(\omega t - \delta)$. So $v(t) = dx/dt = -\omega A\sin(\omega t - \delta)$. $P = Fv = -A\omega F_0\cos(\omega t)\sin(\omega t - \delta)$.

(b) Perform the appropriate substitution which gives the expression quoted in the problem statement.

(c) $\int\cos\theta\sin\theta\,d\theta = \frac{1}{2}\sin^2\theta$; for the limits $\theta = 0$ and $\theta = 2\pi$, this gives zero.

$\cos^2\theta = \frac{1}{2}[1 + \cos(2\theta)]$. The average of $\cos(2\theta) = 0$ over a complete cycle, so $<\cos^2\theta> = \frac{1}{2}$ and

$P_{av} = \frac{1}{2}A\omega F_0\sin\delta$

(d) Note that the hypotenuse of this triangle is the expression in the denominator, which gives the first equation. We can use Equ. 14-49 to reduce this result to the simpler form $\sin\delta = bA\omega/F_0$.

(e) From the above, $A\omega = (F_0\sin\delta)/b$. Thus $P_{av} = \frac{1}{2}F_0^2\sin^2\delta/b = \frac{1}{2}\{(bF_0^2\omega^2)/[m^2(\omega_0^2 - \omega^2)^2 + b^2\omega^2]\}$.

CHAPTER **15**

Wave Motion

1* • A rope hangs vertically from the ceiling. Do waves on the rope move faster, slower, or at the same speed as they move from bottom to top? Explain.

They move faster as they move up because the tension increases due to the weight of the rope below.

5* • Transverse waves travel at 150 m/s on a wire of length 80 cm that is under a tension of 550 N. What is the mass of the wire?

From Equ. 15-3, $\mu = F/v^2$; $m = \mu L = FL/v^2$ $m = (550 \times 0.8/150^2)$ kg = 0.019

9* •• A common method for estimating the distance to a lightning flash is to begin counting when the flash is observed and continue until the thunder clap is heard. The number of seconds counted is then divided by 3 to get the distance in kilometers. (*a*) What is the velocity of sound in kilometers per second? (*b*) How accurate is this procedure? (*c*) Is a correction for the time it takes for the light to reach you important? (The speed of light is 3×10^8 m/s.)

(*a*) $v = 340$ m/s = 0.340 km/s.

(*b*) $s = vt = 0.340t \approx t/3 = 0.333t$; error = 7/340 = 2%.

(*c*) No; e.g., if $t = 3$ s, $s \approx 1$ km, and the time required for light to travel 1 km is only 10^{-5} s.

13* •• (*a*) Compute the derivative of the velocity of sound with respect to the absolute temperature, and show that the differentials dv and dT obey $dv/v = \frac{1}{2}dT/T$. (*b*) Use this result to compute the percentage change in the velocity of sound when the temperature changes from 0 to 27°C. (*c*) If the speed of sound is 331 m/s at 0°C, what is it (approximately) at 27°C? How does this approximation compare with the result of an exact calculation?

(*a*) Follow the same procedure as in Problem 12. Since $v \propto \sqrt{T}$, $dv/v = \frac{1}{2}dT/T$.

(*b*) $\Delta T/T = 27/273$; $\Delta v/v = 4.95\%$.

(*c*) $v_{300} = v_{273}(1.0495) = 347$ m/s. Using the fact that $v \propto \sqrt{T}$ we obtain $v_{300} = 331\sqrt{300/273}$ m/s = 347 m/s.

17* • Show explicitly that the following functions satisfy the wave equation: (*a*) $y(x,t) = k(x + vt)^3$; (*b*) $y(x,t) = Ae^{ik(x-vt)}$, where A and k are constants and $i = \sqrt{-1}$; and (*c*) $y(x,t) = \ln k(x - vt)$.

(*a*) $\partial y/\partial x = 3k(x + vt)^2$; $\partial^2 y/\partial x^2 = 6k(x + vt)$; $\partial y/\partial t = 3kv(x + vt)^2$; $\partial^2 y/\partial t^2 = 6kv^2(x + vt)$. $v^2(\partial^2 y/\partial x^2) = \partial^2 y/\partial t^2$.

(*b*) $\partial y/\partial x = ikAe^{ik(x-vt)}$; $\partial^2 y/\partial x^2 = -k^2 Ae^{ik(x-vt)}$; $\partial y/\partial t = -ikvAe^{ik(x-vt)}$; $\partial^2 y/\partial t^2 = -k^2 v^2 Ae^{ik(x-vt)} = v^2(\partial^2 y/\partial x^2)$.

(c) $\partial y/\partial x = 1/(x - vt)$; $\partial^2 y/\partial x^2 = -1/(x - vt)^2$; $\partial y/\partial t = -v/(x - vt)$; $\partial^2 y/\partial t^2 = -v^2/(x - vt)^2 = v^2(\partial^2 y/\partial x^2)$.

21* • True or false: The energy in a wave is proportional to the square of the amplitude of the wave.

True; see Equ. 15-24.

25* • Equation 15-10 applies to all types of periodic waves, including electromagnetic waves such as light waves and microwaves, which travel at 3×10^8 m/s in a vacuum. (a) The range of wavelengths of light to which the eye is sensitive is about 4×10^{-7} to 7×10^{-7}m. What are the frequencies that correspond to these wavelengths? (b) Find the frequency of a microwave that has a wavelength of 3 cm.

(a) $f = v/\lambda$; $v = c = 3 \times 10^8$ m/s $\qquad$ $f_{min} = (3 \times 10^8/7 \times 10^{-7})$ Hz $\approx 4.3 \times 10^{14}$ Hz;

$\qquad\qquad\qquad\qquad\qquad\qquad\qquad\qquad\qquad$ $f_{max} = 7.5 \times 10^{14}$ Hz

(b) $f = c/\lambda$ $\qquad\qquad\qquad\qquad\qquad\qquad$ $f = 3 \times 10^8/3 \times 10^{-2}$ Hz $= 10^{10}$ Hz

29* •• A harmonic wave with a frequency of 80 Hz and an amplitude of 0.025 m travels along a string to the right with a speed of 12 m/s. (a) Write a suitable wave function for this wave. (b) Find the maximum speed of a point on the string. (c) Find the maximum acceleration of a point on the string.

(a) See Problem 24(e). $y(x, t) = 0.025 \sin [(160\pi)(x/12 - t)]$ m $= 0.025 \sin (41.9x - 503t)$ m.

(b) $v_{max} = A\omega = (0.025 \times 503)$ m/s $= 12.6$ m/s.

(c) $a_{max} = A\omega^2 = \omega v_{max} = 6321$ m/s^2.

33* • A sound wave in air produces a pressure variation given by

$$p(x,t) = 0.75\cos \frac{\pi}{2}(x - 340t)$$

where p is in pascals, x is in meters, and t is in seconds. Find (a) the pressure amplitude of the sound wave, (b) the wavelength, (c) the frequency, and (d) the speed?

(a) $p_0 = 0.75$ Pa. (b) From Problem 24(d), we see that $2\pi/\lambda = \pi/2$, $\lambda = 4$ m. (c) $f = v/\lambda = 85$ Hz. (d) $v = 340$ m/s.

37* •• A typical loud sound wave with a frequency of 1 kHz has a pressure amplitude of about 10^{-4} atm. (a) At $t = 0$, the pressure is a maximum at some point x_1. What is the displacement at that point at $t = 0$? (b) What is the maximum value of the displacement at any time and place? (Take the density of air to be 1.29 kg/m^3.)

(a) When p is a maximum, $s = 0$. (b) $s_0 = p_0/\rho\omega v = 3.67 \times 10^{-6}$ m.

41* • A loudspeaker at a rock concert generates 10^{-2} W/m^2 at 20 m at a frequency of 1 kHz. Assume that the speaker spreads its energy uniformly in three dimensions. (a) What is the total acoustic power output of the speaker? (b) At what distance will the intensity be at the pain threshold of 1 W/m^2? (c) What is the intensity at 30 m?

(a) $P = 4\pi r^2 I = 4\pi \times 400 \times 10^{-2}$ W $= 50.3$ W. (b) Since $I \propto 1/r^2$, r at pain threshold is 2.0 m.

(c) $I = 10^{-2}(4/9) = 4.44 \times 10^{-3}$ W/m^2.

45* • Find the intensity of a sound wave if (a) $\beta = 10$ dB, and (b) $\beta = 3$ dB. (c) Find the pressure amplitudes of sound waves in air for each of these intensities.

(a), (b) Use Equ. 15-29 $\qquad\qquad\qquad$ (a) $I = 10^{-11}$ W/m^2; (b) $I = 2 \times 10^{-12}$ W/m^2

(c) $p_0 = \sqrt{2I\rho v}$ $\qquad\qquad\qquad\qquad$ (a) $p_0 = 9.37 \times 10^{-5}$ Pa; (b) $p_0 = 4.19 \times 10^{-5}$ Pa

49*• What fraction of the acoustic power of a noise would have to be eliminated to lower its sound intensity level from 90 to 70 dB?

99% must be eliminated so that the power is reduced by factor of 100.

53* •• A loudspeaker at a rock concert generates 10^{-2} W/m^2 at 20 m at a frequency of 1 kHz. Assume that the speaker spreads its energy uniformly in all directions. (*a*) What is the intensity level at 20 m? (*b*) What is the total acoustic power output of the speaker? (*c*) At what distance will the intensity level be at the pain threshold of 120 dB? (*d*) What is the intensity level at 30 m?

(*a*) $\beta = 100$ dB at 20 m. (*b*), (*c*), (*d*) See Problem 41; (*d*) 4.44×10^{-3} W/m^2 = 96.5 dB.

57* ••• Everyone at a party is talking equally loudly. If only one person were talking, the sound level would be 72 dB. Find the sound level when all 38 people are talking.

$I_{tot} = 38I_1$; $\beta_{tot} = [10 \log (38) + 72]$ dB = (15.8 + 72) dB = 88.8 dB.

61* • The frequency of a car horn is f_0. What frequency is observed if both the car and the observer are at rest, but a wind blows toward the observer?

(*a*) f_0

(*b*) Greater than f_0

(*c*) Less than f_0

(*d*) It could be either greater or less than f_0.

(*e*) It could be f_0 or greater than f_0, depending on how wind speed compares to speed of sound.

(*a*) There is no relative motion of the source and receiver.

In Problems 65 through 70, a source emits sounds of frequency 200 Hz that travel through still air at 340 m/s.

65* • The sound source described above moves with a speed of 80 m/s relative to still air toward a stationary listener. (*a*) Find the wavelength of the sound between the source and the listener. (*b*) Find the frequency heard by the listener.

(*a*) Apply Equ. 15-31 $\lambda' = (260/200)$ m = 1.3 m

(*b*) Apply Equ. 15-35 $f' = 200(340/260)$ Hz = 262 Hz

69* • Consider the situation in Problem 68 in a reference frame in which the listener is at rest. (*a*) What is the wind velocity in this frame? (*b*) What is the speed of the sound from the source to the listener in this frame, that is, relative to the listener? (*c*) Find the wavelength of the sound between the source and the listener in this frame. (*d*) Find the frequency heard by the listener.

(*a*) Moving at 80 m/s in still air, the observer experiences a wind of 80 m/s.

(*b*) Using the standard Galilean transformation, $v' = v + u_r = 420$ m/s.

(*c*) The distance between wave crests is unchanged, so $\lambda' = \lambda = 1.7$ m.

(*d*) $f' = v'/\lambda' = 247$ Hz.

73* •• A radar device emits microwaves with a frequency of 2.00 GHz. When the waves are reflected from a car moving directly away from the emitter, a frequency difference of 293 Hz is detected. Find the speed of the car.

1. The frequency f received by the car is given by Equ. 15-37*a*.

2. The car now acts as the source, sending signals of frequency f to the stationary radar receiver.

3. Consequently, $f_{rec} = \dfrac{1 + v/c}{1 - v/c} f_0 = (1 + 2v/c)f_0$ since $v \ll c$.

4. Solve for v; $v = c\Delta f/2f_0$ $\qquad\qquad$ $v = (3 \times 10^8 \times 293/4 \times 10^9)$ m/s $\doteq 22$ m/s $= 79.2$ km/h

---

77*•• Suppose the police car of Problem 76 is moving in the same direction as the other vehicle at a speed of 60 km/h. What then is the difference in frequency between the emitted and the reflected signals?

Now the relative velocity is 80 km/h $\qquad\qquad$ $\Delta f = (80 \times 7.78/140)$ kHz $= 4.45$ kHz

---

81*•• Two students with vibrating 440-Hz tuning forks walk away from each other with equal speeds. How fast must they walk so that they each hear a frequency of 438 Hz from the other fork?

See Problem 80. In this case, $f' = (1 - 2u/v)f_0$. $\Delta f = 2f_0 u/v$. $u = \Delta f v/2f_0 = (2 \times 340/880)$ m/s $= 0.773$ m/s.

85*•• A balloon driven by a 36-km/h wind emits a sound of 800 Hz as it approaches a tall building. (a) What is the frequency of the sound heard by an observer at the window of this building? (b) What is the frequency of the reflected sound heard by a person riding in the balloon?

The simplest way to approach this problem is to transform to a reference frame in which the balloon is at rest. In that reference frame, the speed of sound is $v = 340$ m/s, and $u_r = 36$ km/h $= 10$ m/s.

(a) Use Equ. 15-34 $\qquad\qquad\qquad\qquad$ $f' = 800(1 + 10/340)$ Hz $= 823.5$ Hz

(b) f' acts as moving source; use Equ. 15-33 $\qquad$ $f' = 823.5/(1 - 10/340)$ Hz $= 848.5$ Hz

---

89*••• Astronomers can deduce the existence of a binary star system even if the two stars cannot be visually resolved by noting an alternating Doppler shift of a spectral line. Suppose that an astronomical observation shows that the source of light is eclipsed once every 18 h. The wavelength of the spectral line observed changes from a maximum of 563 nm to a minimum of 539 nm. Assume that the double star system consists of a very massive, dark object and a relatively light star that radiates the observed spectral line. Use the data to determine the separation between the two objects (assume that the light object is in a circular orbit about the massive one) and the mass of the massive object. (Use the approximation $\Delta f/f_0 \approx v/c$.)

1. Determine the maximum and minimum frequencies $\qquad$ $f_{max} = \dfrac{3 \times 10^8}{5.39 \times 10^{-7}}$ Hz $= 5.566 \times 10^{14}$ Hz;

$\qquad\qquad\qquad\qquad\qquad\qquad\qquad\qquad$ $f_{min} = 5.329$ Hz

2. $f_0 = \frac{1}{2}(f_{max} + f_{min})$ $\qquad\qquad\qquad\qquad$ $f_0 = 5.4475 \times 10^{14}$ Hz

3. $v = c\Delta f/f_0$ $\qquad\qquad\qquad\qquad\qquad\qquad$ $v = 6.526 \times 10^6$ m/s

4. Determine R, the radius of the orbit $\qquad\quad$ $R = vT/2\pi = (6.526 \times 10^6 \times 64800/2\pi)$ m

$\qquad\qquad\qquad\qquad\qquad\qquad\qquad\qquad$ $= 6.73 \times 10^{10}$ m

5. From Equ. 11-15, $M = \dfrac{4\pi^2 R^3}{GT^2}$ $\qquad$ $M = \dfrac{4\pi^2(6.73 \times 10^{10})^3}{6.67 \times 10^{-11}(6.48 \times 10^4)^2}$ kg $= 4.3 \times 10^{34}$ kg

---

93*• True or false: (a) Wave pulses on strings are transverse waves. (b) Sound waves in air are transverse waves of compression and rarefaction. (c) The speed of sound at 20°C is twice that at 5°C.

(a) True (b) False (c) False

97* •• Consider a long line of cars equally spaced by one car length and moving slowly with the same speed. One car suddenly slows to avoid a dog and then speeds up until it is again one car length behind the car ahead. Discuss how the space between cars propagates back along the line. How is this like a wave pulse? Is there any transport of energy? What does the speed of propagation depend on?

The driver of the car behind slows and then speeds up. This gives rise to a longitudinal wave pulse propagating backwards along the line of cars. There is no transport of energy. The speed of propagation of the pulse depends on the length of a car and on the driver's reaction time.

101*• The following wave functions represent traveling waves:

$(a) y_1(x,t) = A \cos k[x + (34 \text{ m/s})t]$,

$(b) y_2(x,t) = Ae^{k[x - (20 \text{ m/s})t]}$,

$(c) y_3(x,t) = BC + \{k[x - (10 \text{ m/s})t]\}^2$,

where x is in meters, t is in seconds, and $A, k, B,$ and C are constants that have the proper units for y to be in meters. Give the direction of propagation and the speed of the wave for each wave function.

(a) Wave propagates to the left ($-x$ direction) at a speed of 34 m/s.

(b) Wave propagates to the right at 20 m/s.

(c) Wave propagates to the right at 10 m/s.

105*•• Ocean waves move toward the beach with a speed of 8.9 m/s and a crest-to-crest separation of 15.0 m. You are in a small boat anchored off shore. (a) What is the frequency of the ocean waves? (b) You now lift anchor and head out to sea at a speed of 15 m/s. What frequency of the waves do you observe?

Given: $\lambda = 15$ m, $v = 8.9$ m/s.

$(a)\ f_0 = v/\lambda$ $f_0 = 8.9/15$ Hz $= 0.593$ Hz

$(b)\ f' = f_0(1 + u_r/v)$ $f' = 0.593(1 + 15/8.9)$ Hz $= 1.59$ Hz

109*•• Find the speed of a car for which the tone of its horn will drop by 10% as it passes you.

Let $\alpha = u_s/v$.

Then we are given $\dfrac{1 + \alpha}{1 - \alpha} = 0.9$; solve for α and u_s $\alpha = 0.1/1.9;\ u_s = 34/1.9$ m/s $= 17.9$ m/s $= 64.4$ km/h

113*•• Two wires of different linear mass densities are soldered together end to end and are then stretched under a tension F (the tension is the same in both wires). The wave speed in the second wire is three times that in the first wire. When a harmonic wave traveling in the first wire is reflected at the junction of the wires, the reflected wave has half the amplitude of the incident wave. (a) If the amplitude of the incident wave is A, what are the amplitudes of the reflected and transmitted waves? (b) Assuming no loss in the wire, what fraction of the incident power is reflected at the junction and what fraction is transmitted? (c) Show that the displacement just to the left of the junction equals that just to the right of the junction.

(a) 1. From Example 15-9, $A^2/v_1 = A_t^2/v_2 + A_r^2/v_1$ $A^2/v_1 = A_t^2/3v_1 + A_r^2/4v_1$

     2. Solve for A_t and A_r $A_t = 3A/2,\ A_r = \frac{1}{2}A$ (given)

$(b)\ P_r = (A_r^2/A^2)P_i;\ P_t = P_i - P_r$ $P_r = P_i/4;\ P_t = 3P_i/4$

$(c)\ A_{\text{left}} = A + A_r$ $A_{\text{left}} = 3A/2 = A_t$

117*•• A tuning fork attached to a stretched wire generates transverse waves. The vibration of the fork is perpendicular to the wire. Its frequency is 400 Hz, and the amplitude of its oscillation is 0.50 mm. The wire has linear mass density of 0.01 kg/m and is under a tension of 1 kN. Assume that there are no reflected waves. (*a*) Find the period and frequency of waves on the wire. (*b*) What is the speed of the waves? (*c*) What are the wavelength and wave number? (*d*) Write a suitable wave function for the waves on the wire. (*e*) Calculate the maximum speed and acceleration of a point on the wire. (*f*) At what average rate must energy be supplied to the fork to keep it oscillating at a steady amplitude?

(*a*) $f = 400$ Hz (given); $T = 1/f = 2.5$ ms. (*b*) $v = \sqrt{F/\mu} = \sqrt{10^5}$ m/s $= 316$ m/s. (*c*) $\lambda = f/v = 79$ cm; $k = 2\pi/\lambda = 7.95$ m^{-1}. (*d*) $y(x, t) = A \sin (kx - \omega t) = [5 \times 10^{-4} \sin (7.95x - 2.51 \times 10^3 t)]$ m.

(*e*) $v_{max} = A\omega = 1.26$ m/s; $a_{max} = A\omega^2 = 3.16 \times 10^3$ m/s^2.

(*f*) $P_{av} = \frac{1}{2}\mu\omega^2 A^2 v = 10^{-2} \times 2\pi^2 \times 16 \times 10^4 \times 25 \times 10^{-8} \times 316$ W $= 2.5$ W.

121* ••• A heavy rope 3 m long is attached to the ceiling and is allowed to hang freely. (*a*) Show that the speed of transverse waves on the rope is independent of its mass and length but does depend on the distance y from the bottom according to the formula $v = \sqrt{gy}$. (*b*) If the bottom end of the rope is given a sudden sideways displacement, how long does it take the resulting wave pulse to go to the ceiling, reflect, and return to the bottom of the rope?

(*a*) The speed is given by $v = \sqrt{F/\mu}$. At a distance y from the bottom, $F = \mu gy$. Thus $v = \sqrt{gy}$.

(*b*) $dy/dt = v = \sqrt{gy}$. The time to travel up and back is two times t, the time to travel from $y = 0$ to $y = 3$ m. We find t by integration.

$$t = \frac{1}{\sqrt{g}}\int_0^3 \frac{dy}{\sqrt{y}} = 2\sqrt{3/9.81} \text{ s} = 1.106 \text{ s}.$$

Thus the total time is 2.21 s.

CHAPTER 16

Superposition and Standing Waves

1* • True or false: (*a*) The waves from two coherent sources that are radiating in phase interfere constructively everywhere in space. (*b*) Two wave sources that are out of phase by 180° are incoherent. (*c*) Interference patterns are observed only for coherent sources.

(*a*) False (*b*) False (*c*) True

5* • Two waves traveling on a string in the same direction both have a frequency of 100 Hz, a wavelength of 2 cm, and an amplitude of 0.02 m. What is the amplitude of the resultant wave if the original waves differ in phase by (*a*) $\pi/6$, and (*b*) $\pi/3$?

(*a*) $A = 2y_0 \cos \frac{1}{2}\delta$ $A = 4 \cos \pi/12 \text{ cm} = 3.86 \text{ cm}$

(*b*) Repeat for $\delta = \pi/3$ $A = 4 \cos \pi/6 \text{ cm} = 3.46 \text{ cm}$

9* • Two sound sources oscillate in phase with a frequency of 100 Hz. At a point 5.00 m from one source and 5.85 m from the other, the amplitude of the sound from each source separately is A. (*a*) What is the phase difference in the sound waves from the two sources at that point? (*b*) What is the amplitude of the resultant wave at that point?

(*a*) Find $\Delta x/\lambda$, then use Equ. 16-9 $\lambda = 3.4 \text{ m}; \Delta x/\lambda = 0.25; \delta = 90°$

(*b*) Apply Equ. 16-6 $A_{res} = 2A \cos 45° = A\sqrt{2}$

13* • Answer the questions of Problem 12 for a point P' for which the distance to the far speaker is 1λ greater than the distance to the near speaker. Assume that the intensity at point P' due to each speaker separately is again I_0.

(*a*) The path difference is λ, so we have constructive interference. $I = 4I_0$.

(*b*) The sources are incoherent and the intensities add; $I = 2I_0$.

(*c*) The phase difference is π; $I = 0$.

17* •• It is thought that the brain determines the direction toward the source of a sound by sensing the phase difference between the sound waves striking the eardrums. A distant source emits sound of frequency 680 Hz. When you are facing directly toward a sound source there should be no phase difference. Estimate the change in phase difference between the sounds received by the ears as you turn from facing directly toward the source through 90°.

The distance between the ears $\approx$ 20 cm. $\lambda = 340/680$ m = 50 cm. So $\delta \approx 2\pi(20/50) = 0.8\pi$ rad.

21* •• Two sound sources radiating in phase at a frequency of 480 Hz interfere such that maxima are heard at angles of 0° and 23° from a line perpendicular to that joining the two sources. Find the separation between the two sources and any other angles at which a maximum intensity will be heard. (Use the result of Problem 20.)

Since 23° = 0.401 rad is not a "small" angle, we cannot use the small angle approximation.

$\Delta x = \lambda = 340/480$ m = 0.708 m = $d \sin 23°$ $d = 0.708/\sin 23°$ m = 1.81 m

If $d \sin \theta = 2\lambda$, there will be another intensity $\sin \theta = 1.416/1.81 = 0.782;\ \theta = 51.5°$

maximum

25* ••• A radio telescope consists of two antennas separated by a distance of 200 m. Both antennas are tuned to a particular frequency, such as 20 MHz. The signals from each antenna are fed into a common amplifier, but one signal first passes through a phase adjuster that delays its phase by a chosen amount so that the telescope can "look" in different directions. When the phase delay is zero, plane radio waves that are incident vertically on the antennas produce signals that add constructively at the amplifier. What should the phase delay be so that signals coming from an angle $\theta = 10°$ with the vertical (in the plane formed by the vertical and the line joining the antennas) will add constructively at the amplifier?

1. Determine λ and the path difference for $\theta = 10°$ $\lambda = (3 \times 10^8/2 \times 10^7)$ m = 15 m

   Note that we can disregard the path of 2λ $\Delta s = (200 \sin 10°)$ m = 34.73 m = $2.315\lambda = (2 + 0.315)\lambda$

2. Determine the phase difference for $\Delta \lambda = 0.315\lambda$ $\delta = 0.315 \times 2\pi = 1.98$ rad = 113.5°

29* • Two tuning forks have frequencies of 256 and 260 Hz. If the forks are vibrating at the same time, what is the beat frequency?

$f_{beat} = \Delta f = 4$ Hz.

33* • Standing waves result from the superposition of two waves of (a) the same amplitude, frequency, and direction of propagation. (b) the same amplitude and frequency and opposite directions of propagation. (c) the same amplitude, slightly different frequency, and the same direction of propagation. (d) the same amplitude, slightly different frequency, and opposite directions of propagation.

(b)

37* • A string fixed at both ends is 3 m long. It resonates in its second harmonic at a frequency of 60 Hz. What is the speed of transverse waves on the string?

1. Determine λ_2 $\lambda_2 = L = 3$ m

2. $v = f_2\lambda_2$ $v = 60 \times 3$ m/s = 180 m/s

41* • A rope 4 m long is fixed at one end; the other end is attached to a light string so that it is free to move. The speed of waves on the rope is 20 m/s. Find the frequency of (a) the fundamental, (b) the second harmonic, and (c) the third harmonic.

(a) Find $\lambda_1 = 4L; f_1 = v/\lambda_1$ $\lambda_1 = 16$ m; $f_1 = 20/16$ Hz = 1.25 Hz

(b) The system does not support a second harmonic $f_2 = 0$

(c) $f_3 = 3f_1$ $f_3 = 3.75$ Hz

45* •• The wave function $y(x, t)$ for a certain standing wave on a string fixed at both ends is given by $y(x,t) = 4.2 \sin 0.20x \cos 300t$, where y and x are in centimeters and t is in seconds. (a) What are the wavelength and frequency of this wave? (b) What is the speed of transverse waves on this string? (c) If the string is vibrating in its fourth harmonic, how long is it?

(a) See Equ. 16-16 $\lambda = 10\pi$ cm $= 31.4$ cm; $f = 300/2\pi$ Hz $= 47.7$ Hz

(b) $v = f\lambda$ $v = 10\pi \times 300/2\pi$ cm/s $= 1500$ cm/s $= 15$ m/s

(c) $\lambda_4 = \lambda_1/4 = 2L/4 = L/2$ $L = 20\pi$ cm $= 62.8$ cm

---

49* •• Three successive resonance frequencies for a certain string are 75, 125, and 175 Hz. (a) Find the ratios of each pair of successive resonance frequencies. (b) How can you tell that these frequencies are for a string fixed at one end only rather than for a string fixed at both ends? (c) What is the fundamental frequency? (d) Which harmonics are these resonance frequencies? (e) If the speed of transverse waves on this string is 400 m/s, find the length of the string.

(a) The ratios are 3/5 and 5/7.   (b)   There are no even harmonics, so the string must be fixed at one end only.

(c) The fundamental frequency is 75/3 Hz $= 25$ Hz. (d) They are the third, fifth, and seventh harmonics. (e) For the fundamental frequency, $\lambda_1 = v/f_1 = 400/25$ m $= 16$ m, and $L = \lambda_1/4 = 4$ m.

53* •• A violin string of length 40 cm and mass 1.2 g has a frequency of 500 Hz when it is vibrating in its fundamental mode. (a) What is the wavelength of the standing wave on the string? (b) What is the tension in the string? (c) Where should you place your finger to increase the frequency to 650 Hz?

(a) $\lambda = 2L$ $\lambda = 80$ cm

(b) $v = f\lambda = \sqrt{FL/m}$; $F = f^2\lambda^2 m/L$ $F = 480$ N

(c) $L_{650} = L_{500}(500/650)$ $L_{650} = 30.77$ cm; place finger 9.23 cm from scroll bridge

---

57* •• A rubber band with an unstretched length of 0.80 m and a mass of 6×10^{-3} kg stretches to 1.20 m when under a tension of 7.60 N. What is the fundamental frequency of oscillation of this band when stretched between two fixed posts 1.20 m apart?

1. Find v; $v = \sqrt{F/\mu}$; $\mu = m/L$ $v = \sqrt{7.6 \times 1.2/6 \times 10^{-3}}$ m/s $= 39$ m/s

2. $f_1 = v/\lambda = v/2L$ $f_1 = 39/2.4$ Hz $= 16.2$ Hz

61* •• A student carries a small oscillator and speaker as she walks very slowly down a long hall. The speaker emits a sound of frequency 680 Hz which is reflected from the walls at each end of the hall. The student notes that as she walks along, the sound intensity she hears passes through successive maxima and minima. What distance must she walk to pass from one maximum to the next?

The distance between successive maxima corresponds to a path difference of λ for the two reflected waves. As she moves a distance d, the path to the nearer wall and back is reduced by $2d$, that to the farther wall and back is increased by $2d$. Thus for a distance d, the path difference is $4d$. Therefore, she moves a distance $d = \lambda/4$ between successive maxima. Since $\lambda = 340/680$ m $= 50$ cm, she moves 12.5 cm.

65* • A tuning fork of frequency f_0 begins vibrating at time $t = 0$ and is stopped after a time interval Δt. The waveform of the sound at some later time is shown as a function of x. Let N be the (approximate) number of cycles in this waveform. (a) How are N, f_0, and Δt related? (b) If Δx is the length in space of this wave packet, what is the wavelength in terms of Δx and N? (c) What is the wave number k in terms of N and Δx? (d) The number N is uncertain by about ±1 cycle. Use Figure 16-29 to explain why. (*e*) Show that the uncertainty in the wave number due to the uncertainty in N is $2\pi/\Delta x$.

(*a*) The number of cycles in the interval Δt is N; hence $\Delta t \approx NT = N/f_0$.

(*b*) There are about N complete wavelengths in Δx; hence $\lambda \approx \Delta x/N$.

(*c*) $k = 2\pi/\lambda = 2\pi N/\Delta x$.

(*d*) N is uncertain because the waveform dies out gradually rather than stopping abruptly at some time; hence, where the pulse starts and stops is not well defined.

(*e*) $\Delta k = 2\pi\Delta N/\Delta x = 2\pi/\Delta x$.

69* •   A musical instrument consists of drinking glasses partially filled with water that are struck with a small mallet. Explain how this works.

When the edges of the glass vibrate, sound waves are produced in the air in the glass. The resonance frequency of the air columns depends on the length of the air column, which depends on how much water is in the glass.

73* ••   The constant γ for helium (and all monatomic gases) is 1.67. If a man inhales helium and then commences to speak, he sounds like Alvin of the Chipmunks. Why?

The pitch is determined in large part by the resonant cavity of the mouth. Since $v_{He} > v_{air}$, the resonance frequency is higher if helium is the gas in the cavity.

77* •   The shortest pipes used in organs are about 7.5 cm long. (*a*) What is the fundamental frequency of a pipe this long that is open at both ends? (*b*) For such a pipe, what is the highest harmonic that is within the audible range (see Problem 43)?

(*a*) $\lambda_1 = 2L; f_1 = v/\lambda_1 = v/2L; f_n = nf_1$ $f_1 = 340/0.15$ Hz = 2267 Hz; $f_9 = 20400$ Hz is about at frequency threshold

81* ••   A string 5 m long that is fixed at one end only is vibrating in its fifth harmonic with a frequency of 400 Hz. The maximum displacement of any segment of the string is 3 cm. (*a*) What is the wavelength of this wave? (*b*) What is the wave number k? (*c*) What is the angular frequency? (*d*) Write the wave function for this standing wave.

(*a*) $\lambda_1 = 4L; \lambda_5 = 4L/5$ $\lambda_5 = 4$ m

(*b*) $k = 2\pi/\lambda$ $k = \pi/2$ m^{-1}

(*c*) $\omega = 2\pi f$ $\omega = 800\pi$ rad/s

(*d*) $y(x, t) = A \sin kx \cos \omega t$ $y(x, t) = 0.03 \sin(\pi x/2) \cos(800\pi t)$

85* ••   In a lecture demonstration of standing waves, a string is attached to a tuning fork that vibrates at 60 Hz and sets up transverse waves of that frequency on the string. The other end of the string passes over a pulley, and the tension is varied by attaching weights to that end. The string has approximate nodes at the tuning fork and at the pulley. (*a*) If the string has a linear mass density of 8 g/m and is 2.5 m long (from the tuning fork to the pulley), what must the tension be for the string to vibrate in its fundamental mode? (*b*) Find the tension necessary for the string to vibrate in its second, third, and fourth harmonic.

(*a*) $v^2 = F/\mu = f^2\lambda^2 = 4f^2L^2; F = 4f^2L^2\mu$ $F = 4 \times 60^2 \times 2.5^2 \times 0.008$ N = 720 N

(*b*) $f_n = nf_1$ and $F \propto f^2$ $F_2 = 4 \times 720$ N = 2880 N; $F_3 = 6480$ N; $F_4 = 11520$ N

89* •• A 50-cm-long wire fixed at both ends vibrates with a fundamental frequency f_0 when the tension is 50 N. If the tension is increased to 60 N, the fundamental frequency increases by 5 Hz, and a further increase in tension to 70 N results in a fundamental frequency of $(f_0 + 7)$ Hz. Determine the mass of the wire.

1. $f \propto \sqrt{F}$

$(f_0 + 5)/f_0 = \sqrt{60/50} = 1.0954$; $f_0 = 52.4$ Hz

2. $f_0 = v/2L = \sqrt{F/4Lm}$; $m = F/4Lf_0^2$

$m = 50/(4 \times 0.5 \times 52.4^2)$ kg $= 9.1$ g

93* •• (a) Show that if the temperature changes by a small amount ΔT, the fundamental frequency of an organ pipe changes by approximately Δf, where $\Delta f/f = \frac{1}{2}\Delta T/T$. (b) Suppose an organ pipe that is closed at one end has a fundamental frequency of 200 Hz when the temperature is 20°C. What will its fundamental frequency be when the temperature is 30°C? (Ignore any change in the length of the pipe due to thermal expansion.)

(a) $f = CT^{\frac{1}{2}}$, where C is a constant; $df/dT = \frac{1}{2}CT^{-\frac{1}{2}} = \frac{1}{2}f/T$; so $df/f = \frac{1}{2}dT/T$ and $\Delta f/f = \frac{1}{2}\Delta T/T$ if $\Delta T \ll T$.

(b) $\Delta f = (200 \times \frac{1}{2} \times 10/293)$ Hz $= 3.41$ Hz; $f_{30} = f_{20} + \Delta f = 203.4$ Hz.

97* ••• Two sources have a phase difference δ_0 that is proportional to time: $\delta_0 = Ct$, where C is a constant. The amplitude of the wave from each source at some point P is A_0. (a) Write the wave functions for each of the two waves at point P, assuming this point to be a distance x_1 from one source and $x_1 + \Delta x$ from the other. (b) Find the resultant wave function, and show that its amplitude is $2A_0 \cos \frac{1}{2}(\delta + \delta_0)$, where δ is the phase difference at P due to the path difference. (c) Sketch the intensity at point P versus time for a zero path difference. (Let I_0 be the intensity due to each wave separately.) What is the time average of the intensity? (d) Make the same sketch for the intensity at a point for which the path difference is $\frac{1}{2}\lambda$.

(a) $y_1 = A_0 \cos(kx_1 - \omega t)$; $y_2 = A_0 \cos(kx_1 - \omega t + k\Delta x + \delta_0) = A_0 \cos[kx_1 - \omega t + (\delta + \delta_0)]$, where $\delta = k\Delta x$.

(b) Use $\cos \alpha + \cos \beta = 2 \cos \frac{1}{2}(\alpha + \beta) \cos \frac{1}{2}(\alpha - \beta)$; $y_{tot} = y_1 + y_2 = 2A_0 \cos \frac{1}{2}(\delta + \delta_0) \cos \frac{1}{2}[kx_1 - \omega t + \frac{1}{2}(\delta + \delta_0)]$. The amplitude of the resultant wave is $2A_0 \cos \frac{1}{2}(\delta + \delta_0)$.

(c) Note that $I \propto A^2$. With $\delta = 0$ and $\delta_0 = Ct$, $I \propto 4A_0^2 \cos^2 \frac{1}{2}Ct$. The average of $\cos^2 \theta$ over a complete period is $\frac{1}{2}$. So the average intensity is $2I_0$. See the figure to the left, below.

(d) If $\Delta x = \frac{1}{2}\lambda$, then at $t = 0$ the two waves interfere destructively. The plot of intensity versus time is shown below in the figure to the right.

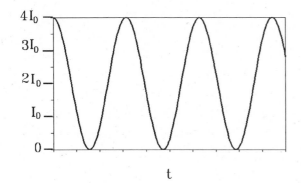

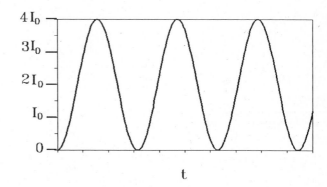

CHAPTER 17

Wave–Particle Duality and Quantum Physics

1* • The quantized character of electromagnetic radiation is revealed by (a) the Young double-slit experiment. (b) diffraction of light by a small aperture. (c) the photoelectric effect. (d) the J.J. Thomson cathode-ray experiment.

(c)

5* • What are the frequencies of photons having the following energies? (a) 1 eV, (b) 1 keV, and (c) 1 MeV.

(a) $f = 1/4.14 \times 10^{-15}$ Hz $= 2.42 \times 10^{14}$ Hz. (b) $f = 2.24 \times 10^{17}$ Hz. (c) $f = 2.24 \times 10^{20}$ Hz.

9* • True or false: In the photoelectric effect, (a) the current is proportional to the intensity of the incident light. (b) the work function of a metal depends on the frequency of the incident light. (c) the maximum kinetic energy of electrons emitted varies linearly with the frequency of the incident light. (d) the energy of a photon is proportional to its frequency.

(a) True (b) False (c) True (d) True

13*• The work function for tungsten is 4.58 eV. (a) Find the threshold frequency and wavelength for the photoelectric effect. (b) Find the maximum kinetic energy of the electrons if the wavelength of the incident light is 200 nm, and (c) 250 nm.

(a) $f_t = \phi/h; \quad \lambda = c/f$ $f_t = 4.58/4.136 \times 10^{-15}$ Hz $= 1.11 \times 10^{15}$ Hz; $\lambda_t = 270$ nm

(b), (c) $K_m = E - \phi = hc/\lambda - \phi$ (b) $K_m = (1240/200 - 4.58)$ eV $= 1.62$ eV

       (c) $K_m = 0.38$ eV

17* •• When a surface is illuminated with light of wavelength 512 nm, the maximum kinetic energy of the emitted electrons is 0.54 eV. What is the maximum kinetic energy if the surface is illuminated with light of wavelength 365 nm?

1. Find $\phi = E - K_m$ $\phi = (1240/512 - 0.54)$ eV $= 1.88$ eV

2. Find K_m for $\lambda = 365$ nm $K_m = (1240/365 - 1.88)$ eV $= 1.52$ eV

21* • Compton used photons of wavelength 0.0711 nm. (a) What is the energy of these photons? (b) What is the wavelength of the photon scattered at $\theta = 180°$? (c) What is the energy of the photon scattered at this angle?

(a) $E = hc/\lambda$ $E = 1240/0.0711$ eV $= 17.44$ keV

(b) Use Equ. 17-8; $\lambda_f = \lambda_i + \Delta\lambda$ $\Delta\lambda = 2 \times 2.43$ pm $= 0.00486$ nm; $\lambda_f = 0.076$ nm

(c) $E = hc/\lambda$ $E = 1240/0.076$ eV $= 16.3$ keV

25* •   True or false: (a) The de Broglie wavelength of an electron varies inversely with its momentum. (b) Electrons can be diffracted. (c) Neutrons can be diffracted. (d) An electron microscope is used to look at electrons.

(a) True (b) True (c) True (d) False

29* •   An electron is moving at $v = 2.5 \times 10^5$ m/s. Find its de Broglie wavelength.

Find $p = mv$; $\lambda = h/p = h/mv$ $\lambda = 6.626 \times 10^{-34}/9.11 \times 10^{-31} \times 2.5 \times 10^5$ m $= 2.91$ nm

---

33* •   Use Equation 17-12 to find the de Broglie wavelength of a proton (rest energy $mc^2 = 938$ MeV) that has a kinetic energy of 2 MeV.

For protons, $\lambda = 2.86 \times 10^{-2}/\sqrt{K}$, λ in nm, K in eV $\lambda = 2.02 \times 10^{-5}$ nm $= 20.2$ fm

---

37* •   The energy of the electron beam in Davisson and Germer's experiment was 54 eV. Calculate the wavelength for these electrons.

Use Equ. 17-13 $\lambda = 0.167$ nm

41* •   Suppose you have a spherical object of mass 4 g moving at 100 m/s. What size aperture is necessary for the object to show diffraction? Show that no common objects would be small enough to squeeze through such an aperture.

For diffraction, the diameter of the aperture $d \approx \lambda$. So $d \approx 6.626 \times 10^{-34}/(4 \times 10^{-3} \times 100) = 1.66 \times 10^{-33}$ m. This is many orders of magnitude smaller than even the diameter of a proton.

45* ••   (a) Find the energy of the ground state ($n = 1$) and the first two excited states of a proton in a one-dimensional box of length $L = 10^{-15}$ m $= 1$ fm. (These are the order of magnitude of nuclear energies.) Make an energy-level diagram for this system and calculate the wavelength of electromagnetic radiation emitted when the proton makes a transition from (b) $n = 2$ to $n = 1$, (c) $n = 3$ to $n = 2$, and (d) $n = 3$ to $n = 1$.

(a) $E_1 = h^2/8mL^2 = 3.28 \times 10^{-11}$ J $= 205$ MeV; the energy level diagram is shown

(b) For $n = 2$ to $n = 1$, $\Delta E = 3E_1$ so $\lambda = 1240/615 \times 10^6$ nm $= 2.02$ fm

(c) For $n = 3$ to $n = 2$, $\Delta E = 5E_1$ and $\lambda = 3 \times 2.02/5$ fm $= 1.21$ fm

(d) For $n = 3$ to $n = 1$, $\Delta E = 8E_1$ and $\lambda = 3 \times 2.02/8$ fm $= 0.758$ fm

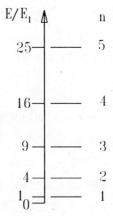

49* ••   Do Problem 48 for a particle in the first excited state ($n = 2$).

$P(x)\Delta x = \psi^2(x)\Delta x$; $\psi(x) = \psi_2(x) = (2/L)^{1/2} \sin(2\pi x/L)$ $P(x) = (2/L) \sin^2(2\pi x/L)$

(a) Evaluate $P(x)$ at $x = L/2$; $P = P(x)\Delta x$ $P = (2/L)(0)(0.002L) = 0$

(b) Repeat as in (a) for $x = 2L/3$ $P = (2/L)(0.75)(0.002L) = 0.003$

(c) Repeat as in (a) for $x = L$ $P = 0$

53* ••   (a) Find $\langle x \rangle$ for the second excited state ($n = 3$) for a particle in a box of length L, and (b) find $\langle x^2 \rangle$.

We proceed as in the preceding problem. Now the integrals over θ extend from 0 to 3π.

(a) $\langle x \rangle = L/2$. (b) $\langle x^2 \rangle = (1/3 - 1/18\pi^2)L^2 = 0.328L^2$. (Note that $\langle x^2 \rangle$ approaches the classical value 1/3 as the quantum number n increases.)

57* ••   Repeat Problem 56 for a particle in the first excited state of the box.

For the first excited state, i.e., for $\psi^2(x) = (2/L) \sin^2 (2\pi x/L)$, the integrals over θ go from 0 to π, 0 to $2\pi/3$, and 0 to $3\pi/2$ for parts (a), (b), and (c), respectively. The other change is that the factor (L/π) is replaced by $(L/2\pi)$.

(a) $P = \dfrac{2}{L} \dfrac{L}{2\pi} \displaystyle\int_0^{\pi} \sin^2 \theta \, d\theta = \dfrac{1}{2}$.

(b) $P = \dfrac{1}{\pi} \displaystyle\int_0^{2\pi/3} \sin^2 \theta \, d\theta = \dfrac{1}{3} + \dfrac{\sqrt{3}}{8\pi} = 0.402$.

(c) $P = \dfrac{1}{\pi} \displaystyle\int_0^{3\pi/2} \sin^2 \theta \, d\theta = \dfrac{3}{4} = 0.75$.

61* •   Can the expectation value of x ever equal a value that has zero probability of being measured?

Yes

65* ••   It was once believed that if two identical experiments are done on identical systems under the same conditions, the results must be identical. Explain why this is not true, and how it can be modified so that it is consistent with quantum physics.

According to quantum theory, the average value of many measurements of the same quantity will yield the expectation value of that quantity. However, any single measurement may differ from the expectation value.

69* •   In 1987, a laser at Los Alamos National Laboratory produced a flash that lasted 1×10^{-12} s and had a power of 5.0×10^{15} W. Estimate the number of emitted photons if their wavelength was 400 nm.

$N = E/E_{ph} = (P\Delta t)/(hc/\lambda)$ $N = (5 \times 10^3 \times 1.6 \times 10^{-19} \text{ eV})/3.1 \text{ eV} = 10^{22}$

73* ••   Suppose that a 100-W source radiates light of wavelength 600 nm uniformly in all directions and that the eye can detect this light if only 20 photons per second enter a dark-adapted eye having a pupil 7 mm in diameter. How far from the source can the light be detected under these rather extreme conditions?

1. At a distance R from the source, the fraction of the light energy entering the eye is $A_{eye}/4\pi R^2 = r^2/4R^2$.

2. Find the number of photons emitted per second $N = P/E_{ph} = 100/[(1240/600) \times 1.6 \times 10^{-19}]$
                                                  $= 3.02 \times 10^{20}$/s

3. Solve for R from $20 = 3.02 \times 10^{20} \times r^2/4R^2$ $R = 6800$ km (neglects absorption by atmosphere)

77* ••   When light of wavelength λ_1 is incident on the cathode of a photoelectric tube, the maximum kinetic energy of the emitted electrons is 1.8 eV. If the wavelength is reduced to $\lambda_1/2$, the maximum kinetic energy of the emitted electrons is 5.5 eV. Find the work function ϕ of the cathode material.

1. Use Equ. 17-3 for λ_1 and $\lambda_1/2$	1.8 eV = $1240/\lambda_1$ – ϕ; 5.5 eV = $2480/\lambda_1$ – ϕ
2. Solve for ϕ	$\phi = 1.9$ eV

81* ••   When a surface is illuminated with light of wavelength λ the maximum kinetic energy of the emitted electrons is 1.2 eV. If the wavelength $\lambda' = 0.8\lambda$ is used the maximum kinetic energy increases to 1.76 eV, and for wavelength $\lambda'' = 0.6\lambda$ the maximum kinetic energy of the emitted electrons is 2.676 eV. Determine the work function of the surface and the wavelength λ.

1. Use Equ. 17-3	$1240/\lambda = 1.2$ eV + ϕ; $1240/0.8\lambda = 1.76$ eV + ϕ
2. Solve for λ	$(1550 - 1240)/\lambda = 310/\lambda = 0.56$ eV; $\lambda = 553.6$ nm
3. Evaluate ϕ	$\phi = 1.04$ eV

85* ••   This problem is one of estimating the time lag (expected classically but not observed) in the photoelectric effect. Let the intensity of the incident radiation be 0.01 W/m². (*a*) If the area of the atom is 0.01 nm², find the energy per second falling on an atom. (*b*) If the work function is 2 eV, how long would it take classically for this much energy to fall on one atom?

(*a*) $P = IA$	$P = 10^{-2} \times 10^{-20}$ J/s = 6.25×10^{-4} eV/s
(*b*) $t = E/P$	$t = 2/6.25 \times 10^{-4}$ s = 3200 s = 53.3 min

---

CHAPTER 18

Temperature and the Kinetic Theory of Gases

1* • True or false:

(*a*) Two objects in thermal equilibrium with each other must be in thermal equilibrium with a third object.

(*b*) The Fahrenheit and Celsius temperature scales differ only in the choice of the zero temperature.

(*c*) The kelvin is the same size as the Celsius degree.

(*d*) All thermometers give the same result when measuring the temperature of a particular system.

(*a*) False   (*b*) False   (*c*) True   (*d*) False

5* • A certain ski wax is rated for use between –12 and –7°C. What is this temperature range on the Fahrenheit scale?

Convert – 12 and – 7°C to t_F using $t_F = (9/5)t_C + 32$ $t_{F1} = 10.4°F$; $t_{F2} = 19.4°F$; between 10.4°F and 19.4°F

9* • The length of the column of mercury in a thermometer is 4.0 cm when the thermometer is immersed in ice water and 24.0 cm when the thermometer is immersed in boiling water. (*a*) What should the length be at room temperature, 22.0°C? (*b*) If the mercury column is 25.4 cm long when the thermometer is immersed in a chemical solution, what is the temperature of the solution?

(*a*), (*b*) $L = [(20/100)t_C + 4]$ cm   (*a*) $L = 8.4$ cm   (*b*) $t_C = (5 \times 21.4)°C = 107°C$

13* • A constant-volume gas thermometer reads 50 torr at the triple point of water. (*a*) What will the pressure be when the thermometer measures a temperature of 300 K? (*b*) What ideal-gas temperature corresponds to a pressure of 678 torr?

(*a*), (*b*)   Use Equ. 18-4   (*a*) $T = 50(300/273.16)$ torr $= 54.9$ torr   (*b*) $T = 3704$ K

17* • The boiling point of oxygen at one atmosphere is 90.2 K. What is the boiling point of oxygen on the Celsius and Fahrenheit scales?

$t_C = T - 273$; then use Equ. 18-2 $t_C = -182.95°C$; $t_F = -297.3°F$

21* •• Figure 18-15 shows a plot of volume versus temperature for a process that takes an ideal gas from point A to point B. What happens to the pressure of the gas?

The pressure increases.

25* •• A pressure as low as 1×10^{-8} torr can be achieved using an oil diffusion pump. How many molecules are there in 1 cm^3 of a gas at this pressure if its temperature is 300 K?

1. Convert torr to atm and cm^3 to L $\qquad$ $P = 10^{-8} \times 1.316 \times 10^{-3}$ atm $= 1.316 \times 10^{-11}$ atm;

$\qquad\qquad\qquad\qquad\qquad\qquad\qquad\qquad$ $V = 1 \times 10^{-3}$ L

2. Use Equs. 18-12 and 18-13 $\qquad$ $N = 1.316 \times 10^{-14} \times 6.022 \times 10^{23}/0.08206 \times 300$

$\qquad\qquad\qquad\qquad\qquad\qquad\qquad\qquad$ $= 3.22 \times 10^8$

29* •• The boiling point of helium at one atmosphere is 4.2 K. What is the volume occupied by helium gas due to evaporation of 10 g of liquid helium at 1 atm pressure and a temperature of (a) 4.2 K, and (b) 293 K?

(a) 10 g = 2.5 mol; $V = nRT/P$ $\qquad$ $V_{4.2} = 2.5 \times 0.08206 \times 4.2/1$ L $= 0.862$ L

(b) $V_2 = V_1(T_2/T_1)$ $\qquad\qquad\qquad\qquad$ $V_{293} = 0.862 \times 293/4.2 = 60.1$ L

33* ••• A helium balloon is used to lift a load of 110 N. The weight of the balloon's skin is 50 N, and the volume of the balloon when fully inflated is 32 m^3. The temperature of the air is 0°C and the atmospheric pressure is 1 atm. The balloon is inflated with sufficient helium gas so that the net buoyant force on the balloon and its load is 30 N. Neglect changes of temperature with altitude.

(a) How many moles of helium gas are contained in the balloon?

(b) At what altitude will the balloon be fully inflated?

(c) Does the balloon ever reach the altitude at which it is fully inflated?

(d) If the answer to (c) is affirmative, what is the maximum altitude attained by the balloon?

(a) Find V from $F_B = mg + 30$ N $\qquad$ $\rho_{air}Vg = 190$ N $+ \rho_{He}Vg$; $V = 17.38$ m^3

$\qquad$ $n = PV/RT$ $\qquad\qquad\qquad\qquad\qquad$ $n = 1 \times 17.38/0.08206 \times 273 = 776$

(b) Find P for $V = 32$ m^3 $\qquad\qquad$ $P = 17.32/32$ atm $= 0.543$ atm

$\qquad$ Determine h from Problem 13-96 $\qquad$ $0.543 = e^{-h/7.93}$; $h = [7.93 \ln(1/.543)]$ km $= 4.84$ km

(c) Determine F_B at 4.84 km; $\rho_{air} = 1.293 \times 0.543$ $\qquad$ $F_B = 1.293 \times 0.543 \times 32 \times 9.81$ N $= 220.4$ N

$\qquad$ Find $m_{tot}g$ $\qquad\qquad\qquad\qquad\qquad\qquad$ $m_{tot}g = (160 + 0.004 \times 776 \times 9.81)$ N $= 190.5$ N

$\qquad$ Note that F_B at 4.84 km is greater than $m_{tot}g$ $\qquad$ Yes; the balloon will rise above 4.84 km

(d) Find ρ_{air} such that $F_B = 190.5$ N $\qquad$ $\rho_{air} = 190.5/32 \times 9.81$ kg/m^3 $= 0.6068$ kg/m^3

$\qquad$ Note that $P = (0.6068/1.293)$ atm $= e^{-h/7.93}$ $\qquad$ $h = [7.93 \ln(1/0.469)]$ km $= 6.0$ km

37* • A mole of He molecules is in one container and a mole of CH$_4$ molecules is in a second container, both at standard conditions. Which molecules have the greater mean free path?

From Equ. 18-25, it follows that the He atoms have the greater mean free path since the diameter of He is smaller than that of CH$_4$.

41* • Find the rms speed and the average kinetic energy of a hydrogen atom at a temperature of 10^7 K. (At this temperature, which is of the order of the temperature in the interior of a star, the hydrogen is ionized and consists of a single proton.)

Use Equs. 18-22 and 18-23; $M = 10^{-3}$ kg/mol $\qquad$ $K = 1.5kT$ J $= 2.07 \times 10^{16}$ J

$$ v_{rms} = \sqrt{\frac{3 \times 8.314 \times 10^7}{10^{-3}}} \text{ m/s} = 499 \text{ m/s} $$

45* •• Repeat Problem 44 for Jupiter, whose escape velocity is 60 km/s and whose temperature is typically −150°C.

(a) $v_{rms} = \sqrt{\dfrac{3RT}{M}}$; M of $H_2 = 2 \times 10^{-3}$ kg/mol $\qquad v_{rms} = \sqrt{\dfrac{3 \times 8.314 \times 123}{2 \times 10^{-3}}}$ m/s = 1368 m/s

(b) M of $O_2 = 32 \times 10^{-3}$ kg/mol $\qquad\qquad v_{rms} = 1368/4$ m/s = 342 m/s

(c) M of $CO_2 = 44 \times 10^{-3}$ kg/mol $\qquad\quad v_{rms} = 1368/\sqrt{22}$ m/s = 291 m/s

(d) $v_{esc}/5 = 12000$ m/s $\qquad\qquad\qquad\qquad$ H_2, O_2 and CO_2 should all be present.

49* •• Show that $f(v)$ given by Equation 18-37 is maximum when $v = \sqrt{2kT/m}$. *Hint:* Set $df/dv = 0$ and solve for v.

The derivative of f with respect to v is $\dfrac{df}{dv} = \dfrac{4}{\sqrt{\pi}}\left(\dfrac{m}{2kT}\right)^{3/2}\left(2v - \dfrac{mv^3}{kT}\right)$. Set this equal to zero and solve for v. The result is $v = \sqrt{2kT/m}$.

53* • True or false: If the pressure of a gas increases, the temperature must increase.

False

57* • If the temperature of an ideal gas is doubled while maintaining constant pressure, the average speed of the molecules

(a) remains constant.

(b) increases by a factor of 4.

(c) increases by a factor of 2.

(d) increases by a factor of $\sqrt{2}$.

(d)

61* • Two different gases are at the same temperature. What can you say about the rms speeds of the gas molecules? What can you say about the average kinetic energies of the molecules?

The rms speeds are inversely proportional to the square root of the molecular masses. The average kinetic energies of the molecules are the same.

65* • At what temperature will the rms speed of an H_2 molecule equal 331 m/s?

From Equ. 18-23, $T = Mv_{rms}^2/3R$ $\qquad\qquad T = (2 \times 10^{-3} \times 331^2/3 \times 8.314)$ K = 8.79 K

69* •• Water, H_2O, can be converted into H_2 and O_2 gas by electrolysis. How many moles of these gases result from the electrolysis of 2 L of water?

$n(H_2O) \rightarrow n(H_2) + 2n(O_2)$; $M(H_2O) = 18$ $\qquad n(H_2O) = 2000/18 = 111$; $n(H_2) = 111$, $n(O_2) = 55.5$

73* •• Three insulated vessels of equal volumes V are connected by thin tubes that can transfer gas but do not transfer heat. Initially all vessels are filled with the same type of gas at a temperature T_0 and pressure P_0. Then the temperature in the first vessel is doubled and the temperature in the second vessel is tripled. The temperature in the third vessel remains unchanged. Find the final pressure P' in the system in terms of the initial pressure P_0. Initially, we have $3P_0V = n_0RT_0$. Later, the pressures in the three vessels, each of volume V, are still equal, but the number of moles are not. We can now write

$$P' = P_1 = P_2 = P_3 = \frac{n_1 R 2 T_0}{V} = \frac{n_2 R 3 T_0}{V} = \frac{n_3 R T_0}{V}. \quad \text{Also,}$$

$$\sum_{i=1}^{3} n_i = n_0 = \frac{3 P_0 V}{R T_0} = \frac{P' V}{R T_0}\left(\frac{1}{2} + \frac{1}{3} + 1 \right) = \frac{P' V}{R T_0}\left(\frac{11}{6} \right), \quad \text{and solving for } P', \text{ we find} \quad P' = \frac{18}{11} P_0.$$

CHAPTER 19

Heat and the First Law of Thermodynamics

1* • Body A has twice the mass and twice the specific heat of body B. If they are supplied with equal amounts of heat, how do the subsequent changes in their temperatures compare?

$M_A = 2M_B$; $c_A = 2c_B$; $C = Mc$; $\Delta T = Q/C$ $C_A = 4C_B$; $\Delta T_A = \Delta T_B/4$

5* • A solar home contains 10^5 kg of concrete (specific heat = 1.00 kJ/kg·K). How much heat is given off by the concrete when it cools from 25 to 20°C?

$Q = C\Delta T = mc\Delta T$ $Q = (10^5 \times 10^3 \times 5)$ J = 500 MJ

9* •• If 500 g of molten lead at 327°C is poured into a cavity in a large block of ice at 0°C, how much of the ice melts?

$- m_{Pb}(L_{f,Pb} + c_{pb}\Delta T) + m_w L_{f,w} = 0$; solve for m_w $m_w = [500(24.7 + 0.128 \times 327)/333.5]$ g = 99.8 g

---

13* • The specific heat of a certain metal can be determined by measuring the temperature change that occurs when a piece of the metal is heated and then placed in an insulated container made of the same material and containing water. Suppose a piece of metal has a mass of 100 g and is initially at 100°C. The container has a mass of 200 g and contains 500 g of water at an initial temperature of 20.0°C. The final temperature is 21.4°C. What is the specific heat of the metal?

$m_1 c(t_{1i} - t_f) = m_2 c(t_f - t_{2i}) + m_w c_w(t_f - t_{2i})$; find c $78.6c = 2.8c + 7$; $c = 0.093$ cal/g·K

17* •• A well-insulated bucket contains 150 g of ice at 0°C. (*a*) If 20 g of steam at 100°C is injected into the bucket, what is the final equilibrium temperature of the system? (*b*) Is any ice left afterward?

(*a*), (*b*) $m_{st}L_v + m_{st}(100 - t_f) = m_{ice}(L_f + t_f)$; $t_f = 4.97°C$; (*b*) Since $t_f > 0°C$, no ice is left.
solve for t_f

21* ••. Between innings at his weekly softball game, Stan likes to have a sip or two of beer. He usually consumes about 6 cans, which he prefers at exactly 40°F. His wife Bernice puts a six-pack of 12-ounce aluminum cans of beer (1 ounce has a mass of 28.4 g) originally at 80°F in a well-insulated Styrofoam container and begins adding ice. How many 30-g ice cubes must she add to the container so that the final temperature is 40°F? (Neglect heat losses through the container and the heat removed from the aluminum and assume that the beer is mostly water.)

1. Convert to Celsius degrees. $40\,^\circ F = 4.44^\circ C;\ \ 80^\circ F = 26.67^\circ C$

2. $m_B c_B \Delta t_B = n_{ice} m_{ice}(L_f + 4.44)$; solve for n_{ice} $n_{ice} = (6 \times 12 \times 28.4 \times 22.2/30 \times 84.14) = 18$

25* •   Can a system absorb heat with no change in its internal energy?

Yes

29* •   A lead bullet moving at 200 m/s is stopped in a block of wood. Assuming that all of the energy change goes into heating the bullet, find the final temperature of the bullet if its initial temperature is 20°C.

$Q = \tfrac{1}{2}mv^2 = mc\Delta t = mc(t_f - t_i);\ t_f = t_i + v^2/2c$ $t_f = (20 + 200^2/2 \times 128)^\circ C = 176^\circ C$

33* ••   A piece of ice is dropped from a height H. (a) Find the minimum value of H such that the ice melts when it makes an inelastic collision with the ground. Assume that all the mechanical energy lost goes into melting the ice. (b) Is it reasonable to neglect the variation in the acceleration of gravity in doing this problem? (c) Comment on the reasonableness of neglecting air resistance. What effect would air resistance have on your answer?

(a) To melt the ice (at $t = 0^\circ C$), $mgh = mL_f$; $h = L_f/g$ $h = 333.5/9.81$ km = 34 km

(b) Yes. Since $h \ll R_E = 6370$ km, one can neglect the variation of g.

(c) The piece of ice (depending on its mass and shape) will reach its terminal velocity long before striking the ground, and some of the ice will melt before it reaches the ground. However, the relation $\Delta U = mgh = mL_f$ remains valid, so air resistance does not affect h.

37* •   A certain gas consists of ions that repel each other. The gas undergoes a free expansion with no heat exchange and no work done. How does the temperature of the gas change? Why?

The temperature of the gas increases. The average kinetic energy increases with increasing volume due to the repulsive interaction between the ions.

In Problems 39 through 42, the initial state of 1 mol of an ideal gas is $P_1 = 3$ atm, $V_1 = 1$ L, and $U_1 = 456$ J, and its final state is $P_2 = 2$ atm, $V_2 = 3$ L, and $U_2 = 912$ J.

41* ••   The gas is allowed to expand isothermally until its volume is 3 L and its pressure is 1 atm. It is then heated at constant volume until its pressure is 2 atm. (a) Show this process on a PV diagram, and calculate the work done by the gas. (b) Find the heat added during this process.

(a) The path from the initial state I to the final state B is shown on the PV diagram. Here, P is in atmospheres and V in liters. The work done by the gas is equal to the area under the path. Here, the work (area under the curve) is given by Equ. 19-16. We replace nRT_1 by P_1V_1.

$W = 303 \ln(3)$ kJ = 333 kJ.

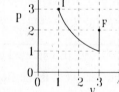

(b) $Q = W + \Delta U$ $Q = (333 + 456)$ J = 789 J

45* ••   One mole of an ideal gas initially at a pressure of 1 atm and a temperature of 0°C is compressed isothermally and quasi-statically until its pressure is 2 atm. Find (a) the work needed to compress the gas, and (b) the heat removed from the gas during the compression.

(a) Use Equ. 19-16, and $V_2/V_1 = P_1/P_2$ W (on the gas) = $8.314 \times 273 \ln(2) = 1573$ J

(b) $\Delta T = 0$, $\Delta U = 0$; $Q = W$ (by the gas) Q (removed from the gas) = 1573 J

49* ••   The specific heat of air at $0°C$ is listed in a handbook as having the value of 1.00 J/g·K measured at constant pressure. (*a*) Assuming that air is an ideal gas with a molar mass $M = 29.0$ g/mol, what is its specific heat at $0°C$ and constant volume? (*b*) How much internal energy is there in 1 L of air at $0°C$ and at 1 atm?

(*a*) For a diatomic gas, $C_v = (5/7)C_p$ 　　　　　　$C_v = 0.714$ J/g·K

(*b*) $\rho_{air} = 1.29$ g/L; $U = C_vT$ 　　　　　　$U = 1.29 \times 0.714 \times 273$ J $= 252$ J

53* ••   The heat capacity of a certain amount of a particular gas at constant pressure is greater than that at constant volume by 29.1 J/K. (*a*) How many moles of the gas are there? (*b*) If the gas is monatomic, what are C_v and C_p? (*c*) If the gas consists of diatomic molecules that rotate but do not vibrate, what are C_v and C_p?

(*a*) $nR = 29.1$ J/K 　　　　　　　　$n = 29.1/8.314 = 3.5$

(*b*) $C_v = 3nR/2$; $C_p = 5nR/2$ 　　　$C_v = 1.5 \times 29.1$ J/K $= 43.65$ J/K; $C_p = 72.75$ J/K

(*c*) $C_v = 5nR/2$; $C_p = 7nR/2$ 　　　$C_v = 72.75$ J/K; $C_p = 101.85$ J/K

57* •   One mole of an ideal gas ($\gamma = \frac{5}{3}$) expands adiabatically and quasi-statically from a pressure of 10 atm and a temperature of $0°C$ to a pressure of 2 atm. Find (*a*) the initial and final volumes, (*b*) the final temperature, and (*c*) the work done by the gas.

(*a*) $V_i = 22.4 \times 1/P_i$ L; from Equ. 19-37, 　$V_i = 2.24$ L; $V_f = 2.24(5)^{0.6}$ L $= 5.88$ L
　　$V_f = V_i(P_i/P_f)^{1/\gamma}$

(*b*) $T_f = P_fV_f/R$ 　　　　　　　$T_f = (202 \times 5.88/8.314)$ K $= 143$ K

(*c*) $W = Q - \Delta U = 0 - C_v\Delta T = -C_v\Delta T$ 　$W = 1.5 \times 8.314 \times 130$ J $= 1.62$ kJ

61* ••   Repeat Problem 60 for a diatomic gas.

(*a*) 1. Find V_i from the ideal gas law. 　　$V_i = (8.314 \times 300/2 \times 400)$ L $= 3.12$ L

2. For the isothermal case, $V_f = V_i(P_i/P_f)$ 　$V_f = 3.12(400/160)$ L $= 7.8$ L

3. $T_f = T_i$ 　　　　　　　　$T_f = 300$ K

4. Use Equ. 19-16; $W = nRT\ln(V_f/V_i)$ 　$W = 0.5 \times 8.314 \times 300 \times \ln(2.5)$ J $= 1.14$ kJ

5. $Q = \Delta U + W$; $\Delta U = 0$ 　　　$Q = 1.14$ kJ

(*b*) 1. From Equ. 19-37, $V_f = V_i(P_i/P_f)^{1/\gamma}$ 　$V_i = 3.12$ L; $V_f = 3.12(2.5)^{0.714} = 6.0$ L

2. $T_f = P_fV_f/nR$ 　　　　　　$T_f = (160 \times 6.0/0.5 \times 8.314)$ K $= 231$ K

3. $W = -\Delta U = -C_v\Delta T$; $Q = 0$ 　$W = (0.5 \times 2.5 \times 8.314 \times 69)$ J $= 717$ J; $Q = 0$

65* ••   One mole of N_2 ($C_v = \frac{5}{2}R$) gas is originally at room temperature ($20°C$) and a pressure of 5 atm. It is allowed to expand adiabatically and quasi-statically until its pressure equals the room pressure of 1 atm. It is then heated at constant pressure until its temperature is again $20°C$. During this heating, the gas expands. After it reaches room temperature, it is heated at constant volume until its pressure is 5 atm. It is then compressed at constant pressure until it is back to its original state. (*a*) Construct an accurate PV diagram showing each process in the cycle. (*b*) From your graph, determine the work done by the gas during the complete cycle. (*c*) How much heat is added or subtracted from the gas during the complete cycle? (*d*) Check your graphical determination of the work done by the gas in (*b*) by calculating the work done during each part of the cycle.

(*a*) 1. Find V at start of cycle, point A, from $V_A = nRT/P$.
　　$V_A = (8.314 \times 293/505)$ L $= 4.82$ L.

2. Find V_B. $V_B = V_A(P_A/P_B)^{1/\gamma} = 4.82(5)^{0.714}$ L = 15.2 L

3. Find $V_C = V_D = (8.314 \times 293/101)$ L = 24.0 L

The complete cycle is shown in the diagram. Here P is in atmospheres and V in liters.

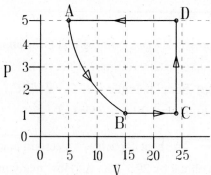

(b) Note that for the paths A-B and B-C, W, the work done by the gas, is positive. For the path D-A, W is negative, and greater in magnitude than W_{A-C}. The total work done by the gas is negative and its magnitude is the area enclosed by the cycle. Each rectangle of the dotted lines equals 5 atm·L. Counting these rectangles, the approximate work done by the gas is about -13×5 atm·L = -65 atm·L.

(c) Since U is a state function, $\Delta U = 0$ for the complete cycle. Consequently, $Q = W = -65$ atm·L = -6.43 kJ

(d) 1. A-B is an adiabatic process. $Q_{A-B} = 0$

2. B-C, $Q = C_p\Delta T$; $C_p = 7R/2$; $T_B = T_A(V_A/V_B)^{\gamma-1}$ $T_B = 293(4.82/15.2)^{0.4}$ K = 185 K; $Q_{B-C} = 3.14$ kJ

3. C-D, $Q = C_v\Delta T$; $C_v = 5R/2$; $T_D = P_D V_D/R$ $T_D = (505 \times 24/8.314)$ K = 1458 K; $Q_{C-D} = 24.2$ kJ

4. D-A, $Q = C_p\Delta T$ $Q_{D-A} = [7 \times 8.314 \times (-1165)/2]$ J = -33.9 kJ

5. $Q_{tot} = Q_{A-B} + Q_{B-C} + Q_{C-D} + Q_{D-A}$ $Q_{tot} = (3.14 + 24.2 - 33.9)$ kJ = -6.54 kJ; fair agreement with -6.43 kJ of part (c).

69* ••• Repeat Problem 67 with a diatomic gas.

1. Find the volume at D from $V = nRT/P$ $V_D = (2 \times 8.314 \times 360/202)$ L = 29.6 L

2. Find V_B, V_C, P_C, and P_B $V_B = V_C = 3V_D = 88.8$ L; $P_C = (2/3)$ atm, $P_B = (4/3)$ atm

3. Find T_C, T_B, and T_A $T_C = T_D = 360$ K; $T_B = T_A = 2T_D = 720$ K

	A	B	C	D
P (atm)	4	4/3	2/3	2
V (L)	29.6	88.8	88.8	29.6
T (K)	720	720	360	360

4. D-A: $W = 0$; $Q = \Delta U = C_v\Delta T$ $W_{D-A} = 0$; $Q_{D-A} = \Delta U_{D-A} = 5 \times 8.314 \times 360$ J = 15.0 kJ

A-B: $W = nRT\ln(V_B/V_A) = Q$; $\Delta U = 0$ $W_{A-B} = 2 \times 8.314 \times 720 \times \ln(3)$ J = 13.16 kJ;

$Q_{A-B} = 13.16$ kJ

B-C: $W = 0$; $Q = \Delta U = C_v\Delta T$ $W_{B-C} = 0$; $Q_{B-C} = -5 \times 8.314 \times 360$ J = -15.0 J

C-D: $W = nRT\ln(V_D/V_C) = Q$; $\Delta U = 0$ $W_{C-D} = -2 \times 8.314 \times 360 \times \ln(3)$ J = -6.58 kJ;

$Q_{C-D} = -6.58$ kJ

5. $W_{tot} = W_{D-A} + W_{A-B} + W_{B-C} + W_{C-D}$ $W_{tot} = 6.58$ kJ

73* •   True or false:

(a) The heat capacity of a body is the amount of heat it can store at a given temperature.

(b) When a system goes from state 1 to state 2, the amount of heat added to the system is the same for all processes.

(c) When a system goes from state 1 to state 2, the work done on the system is the same for all processes.

(d) When a system goes from state 1 to state 2, the change in the internal energy of the system is the same for all processes.

(e) The internal energy of a given amount of an ideal gas depends only on its absolute temperature.

(f) A quasi-static process is one in which there is no motion.

(g) For any material that expands when heated, C_p is greater than C_v.

(a) False. (b) False. (c) False. (d) True. (e) True. (f) False. (g) True.

77* ••   An ideal gas undergoes a process during which $P\sqrt{V}$ = constant and the volume of the gas decreases. What happens to the temperature?

$PV = C\sqrt{V} = nRT$. If V decreases, the T decreases.

81* •   The PV diagram in Figure 19-18 represents 3 mol of an ideal monatomic gas. The gas is initially at point A. The paths AD and BC represent isothermal changes. If the system is brought to point C along the path AEC, find (a) the initial and final temperatures, (b) the work done by the gas, and (c) the heat absorbed by the gas.

Although not required for this problem, we begin by determining pressures, volumes, and temperatures at points A, B, C, D, and E, and then list these as in the table below.

(a) Find $T_A = T_D$ and $T_C = T_B$ using $T = PV/nR$ $\quad T_A = (4 \times 404/3 \times 8.314)$ K = 64.8 K; T_C = 81 K

Find V_B and V_A using $V = nRT/P$ $\quad V_B = (81/64.8) \times 4.0 = 5.0$ L; $V_D = 16.0$ L

	A	B	C	D	E
P (atm)	4.0	4.0	1.0	1.0	1.0
V (L)	4.0	5.0	20.0	16.0	4.0
T (K)	64.8	81.0	81.0	64.8	16.2

(b) $W_{A-E} = 0$; $W_{E-C} = P_E \Delta V$; $\quad W = 16$ atm·L = 1.62 kJ

(c) $Q = W + \Delta U$; $\Delta U = C_v \Delta T$; $C_v = 3nR/2$ $\quad Q = (1.62 + 9 \times 8.314 \times 16.2/2)$ kJ = 2.23 kJ

85* ••   Repeat Problem 84 for the path ADC.

(a) T is a state function. See Problem 81 for T_A, T_C $\quad T_A = 64.8$ K; $T_C = 81.0$ K

(b) 1. Find V_D using PV^γ = constant; $\gamma = 5/3$ $\quad V_D = 4 \times 4^{0.6}$ L = 9.19 L

2. Find T_D using the ideal gas law. $\quad T_D = 37.2$ K

3. $W_{A-D} = -C_v \Delta T_{AD}$; $W_{D-C} = P_D \Delta V_{D-C}$; $\quad W_{A-D} = 9 \times 8.314 \times 27.6/2$ J = 1.03 kJ;

$W = W_{A-D} + W_{D-C}$ $\quad W_{D-C} = 0.101 \times 10.81$ kJ = 1.09 kJ; $W = 2.12$ kJ

(c) $Q = W + \Delta U$; $\Delta U = 0.61$ kJ (see Problem 81) $\quad Q = 2.73$ kJ

89* ••   Repeat Problem 87 with the diatomic ideal gas replaced by a monatomic ideal gas.

1. $W = nRT \ln(V_f/V_i) = Q + \Delta U$; $\Delta U = 0$ $\quad W = -170$ cal = -711 J; W (on gas) = 711 J

2. $\Delta U = 0$ in an isothermal process. $\quad \Delta U = 0$

3. $T_i = T_f = W/[nR \ln(V_f/V_i)]$ $\quad T_i = T_f = -711/[2 \times 8.314 \times \ln(8/18)]$ K = 52.7 K

93* ••   Repeat Problem 91 if the gas is argon.

For Ar, $M = 40$, so $n = 0.75$; $\gamma = 1.67$. Following the procedure of the two preceding problems we obtain:

$V_i = 16.8$ L and $V_f = 3.36$ L. For (a) W(on gas) = 2.74 kJ; for (b) W(on gas) = 48.5 atm·L = 4.9 kJ.

97* ••   Heat in the amount of 500 J is supplied to 2 mol of an ideal diatomic gas. (a) Find the change in temperature if the pressure is kept constant. (b) Find the work done by the gas. (c) Find the ratio of the final volume of the gas to the initial volume if the initial temperature is 20°C.

(a) $\Delta T = Q/C_p$; $C_p = (7/2)nR$ $\Delta T = (500/7 \times 8.314)$ K = 8.59 K

(b) $W = Q - \Delta U = Q - (5/2)nR\Delta T = nR\Delta T$ $W = 2 \times 8.314 \times 8.59$ J = 143 J

(c) $V_f/V_i = T_f/T_i = (T_i + \Delta T)/T_i$ $V_f/V_i = 281.74/273.15 = 1.03$

101* •• One mole of monatomic gas, initially at temperature T, undergoes a process in which its temperature is quadrupled and its volume is halved. Find the amount of heat Q transferred to the gas. It is known that in this process the pressure was never less than the initial pressure, and the work done on the gas was the minimum possible.

The path for this process is shown on the PV diagram. Since $P_fV_f = 4P_iV_i$ and $V_f = V_i/2$, the path for which the work done by the gas is a minimum while the pressure never falls below P_i is shown on the adjacent PV diagram. We can now determine W and ΔU in terms of the initial temperature T, initial pressure P_i, and initial volume V_i.

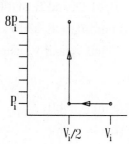

$W = -P_iV_i/2 = -RT/2$. $\Delta U = C_v\Delta T = (3/2)R(3T) = 9RT/2$. $Q = W + \Delta U = 4RT$.

105* ••• Prove that the slope of the adiabatic curve passing through a point on the PV diagram for an ideal gas is γ times the slope of the isothermal curve passing through the same point.

The slope of the curve on a PV diagram is dP/dV. 1. For an isothermal process, $PV = $ constant $= C$. So, $P = C/V$, and $dP/dV = -C/V^2 = -P/V$. 2. For an adiabatic process, $PV^\gamma = C$, and $dP/dV = -\gamma P/V$. We see that the slope for the adiabatic process is steeper by the factor γ.

Note: Problems 106 through 109 involve non-quasi-static processes. Nevertheless, assuming that the gases participating in these processes approximate ideal gases, one can calculate the state functions of the end products of the reactions using the first law of thermodynamics and the ideal gas law. For T > 2000 K, vibration of the atoms contributes to c_p of H_2O and CO_2 so that c_p of these gases is 7.5R at high temperatures. Also, assume the gases do not dissociate.

109* ••• Suppose that instead of pure oxygen, just enough air is mixed with the two mol of CO in the container of Problem 108 to permit complete combustion. Air is 80% N_2 and 20% O_2 by weight, and the nitrogen does not participate in the reaction. What then are the answers to parts (a), (b), and (c) of Problem 108?

Note that for N_2, c_v at temperatures above 2000 K is $(5/2)R + R$ since there is only one vibrational mode that contributes to the specific heat.

(a) 1. Write the reaction for 2 mol of CO $2(CO) + O_2 + 4N_2 \rightarrow 2(CO_2) + 4N_2$

    2. Find P_i of 7 mol at 300 K in 80 L $P_i = (7 \times 8.314 \times 300/80)$ kPa = 218.2 kPa

(b) 1. Find C_v of product gases for $T < 2000$ K $C_v = [2 \times (7/2) + 4 \times (5/2)]R = 141.3$ J/K

2. Find Q to heat gases to 2000 K

$Q = 1700 \times 141.3$ J $= 240.2$ kJ

3. Find Q available to raise gases above 2000 K

$Q = (560 - 240.2)$ kJ $= 319.8$ kJ

4. Find T_f; note that $C_v = 2 \times 6.5$

$\Delta T = (319.8 \times 10^3/27 \times 8.314)$ K $= 1425$ K; $T_f = 3425$ K

5. Find $P_f = P_i(n_f/n_i)(T_f/T_i)$

$P_f = 218.2(6/7)(3425/300)$ kPa $= 2.135$ MPa

(c) $P_f = P_i(n_f/n_i)(T_f/T_i)$

$P_f = 2135(273/3425)$ kPa $= 170.2$ kPa

CHAPTER 20

The Second Law of Thermodynamics

1* • Where does the energy come from in an internal-combustion engine? In a steam engine?

Internal combustion engine: From the heat of combustion (see Problems 19-106 to 19-109).

Steam engine: From the burning of fuel to evaporate water and to raise the temperature and pressure of the steam.

5* • An engine with 20% efficiency does 100 J of work in each cycle. (*a*) How much heat is absorbed in each cycle? (*b*) How much heat is rejected in each cycle?

(*a*) From Equ. 20-2, $Q_h = W/\varepsilon$ $\qquad\qquad$ $Q_h = 100/0.2$ J = 500 J

(*b*) $|Q_c| = Q_h(1- \varepsilon)$ $\qquad\qquad$ $|Q_c| = 500 \times 0.8$ J = 400 J

9* •• An engine operates with 1 mol of an ideal gas for which $C_v = \frac{3}{2}R$ and $C_p = \frac{5}{2}R$ as its working substance. The cycle begins at $P_1 = 1$ atm and $V_1 = 24.6$ L. The gas is heated at constant volume to $P_2 = 2$ atm. It then expands at constant pressure until $V_2 = 49.2$ L. During these two steps, heat is absorbed by the gas. The gas is then cooled at constant volume until its pressure is again 1 atm. It is then compressed at constant pressure to its original state. During the last two steps, heat is rejected by the gas. All the steps are quasi-static and reversible. (*a*) Show this cycle on a *PV* diagram. Find the work done, the heat added, and the change in the internal energy of the gas for each step of the cycle. (*b*) Find the efficiency of the cycle.

(*a*) The cycle is shown on the right. Here, the pressure *P* is in atm and the volume *V* is in L. To determine the heat added during each step we shall first find the temperatures at points 1, 2, 3, and 4.

$T_1 = 24.6 \times 273/22.4$ K = 300 K

$T_2 = 2T_1 = 600$ K

$T_3 = 2T_2 = 1200$ K

$T_4 = 2T_1 = 600$ K

$W_{1\text{-}2} = P\Delta V_{1\text{-}2};\ Q_{1\text{-}2} = \Delta U_{12} = C_v\Delta T_{1\text{-}2}$

$W_{2\text{-}3} = P\Delta V_{2\text{-}3};\ Q_{2\text{-}3} = C_p\Delta T_{2\text{-}3}$

$W_{1\text{-}2} = 0;\ Q_{1\text{-}2} = 1.5 \times 8.314 \times 300$ J = 3.74 kJ = $\Delta U_{1\text{-}2}$

$W_{2\text{-}3} = 2 \times 24.6$ atm·L = 4.97 kJ;

$Q_{2\text{-}3} = 2.5 \times 8.314 \times 600$ J = 12.47 kJ;

$W_{3\text{-}4} = P\Delta V_{3\text{-}4}; \; Q_{3\text{-}4} = C_v\Delta T_{3\text{-}4}$

$W_{4\text{-}1} = P\Delta V_{4\text{-}1}; \; Q_{4\text{-}1} = C_p\Delta T_{4\text{-}1}$

$\Delta U_{2\text{-}3} = (12.47 - 4.97)\text{ kJ} = 7.5\text{ kJ}$

$W_{3\text{-}4} = 0; \; Q_{34} = -1.5 \times 8.314 \times 600\text{ J} = -7.48\text{ kJ} = \Delta U_{3\text{-}4}$

$W_{4\text{-}1} = -24.6\text{ atm·L} = -2.48\text{ kJ}; \; Q_{4\text{-}1} = -6.24\text{ kJ};$

$\Delta U_{4\text{-}1} = -3.76\text{ kJ}$

(b) $\varepsilon = W/Q_{in}$

$W = 2.48\text{ kJ}; \; Q_{in} = 16.21\text{ kJ}. \; \varepsilon = 0.153 = 15.3\%$

13* •• An ideal gas ($\gamma = 1.4$) follows the cycle shown in Figure 20-12. The temperature of state 1 is 200 K. Find (a) the temperatures of the other three states of the cycle and (b) the efficiency of the cycle.

(a) Use $PV = nRT$; $T_i = T_1(P_iV_i/P_1V_1)$ $T_1 = 200$ K, $T_2 = 600$ K, $T_3 = 1800$ K, $T_4 = 600$ K

(b) Find W = area enclosed by cycle. $W = 400$ atm·L

Find $Q_{in} = C_v\Delta T_{1\text{-}2} + C_p\Delta T_{2\text{-}3}$ $Q_{in} = (2.5 \times 200 + 3.5 \times 600)\text{ atm·L} = 2600\text{ atm·L}$

$\varepsilon = W/Q_{in}$ $\varepsilon = 400/2600 = 0.154 = 15.4\%$

17* •• A certain engine running at 30% efficiency draws 200 J of heat from a hot reservoir. Assume that the refrigerator statement of the second law of thermodynamics is false, and show how this engine combined with a perfect refrigerator can violate the heat-engine statement of the second law.

For this engine, $Q_h = 200$ J, $W = 60$ J, and $Q_c = -140$ J. A "perfect" refrigerator would transfer 140 J from the cold reservoir to the hot reservoir with no other effects. Running the heat engine connected to the perfect refrigerator would then have the effect of doing 60 J of work while taking 60 J of heat from the hot reservoir without rejecting any heat, in violation of the heat-engine statement of the second law.

21* • A refrigerator works between an inside temperature of 0°C and a room temperature of 20°C. (a) What is the largest possible coefficient of performance it can have? (b) If the inside of the refrigerator is to be cooled to −10°C, what is the largest possible coefficient of performance it can have, assuming the same room temperature of 20°C?

(a) Express the COP in terms of T_h and T_c. $\text{COP} = |Q_c|/W = |Q_c|/\varepsilon Q_h = (1 - \varepsilon)/\varepsilon = T_c/(T_h - T_c)$

Evaluate COP $\text{COP} = 273/20 = 13.7$

(b) Evaluate COP $\text{COP} = 263/30 = 8.77$

25* •• A Carnot engine works between two heat reservoirs as a refrigerator. It does 50 J of work to remove 100 J from the cold reservoir and gives off 150 J to the hot reservoir during each cycle. Its coefficient of performance $\text{COP} = Q_c/W = (100\text{ J})/(50\text{ J}) = 2$. (a) What is the efficiency of the Carnot engine when it works as a heat engine between the same two reservoirs? (b) Show that no other engine working as a refrigerator between the same two reservoirs can have a COP greater than 2.

(a) The efficiency is given by $\varepsilon = W/Q_h$; $\varepsilon = 50/150 = 0.333 = 33.3\%$.

(b) If COP > 2, then 50 J of work will remove more than 100 J of heat from the cold reservoir and put more than 150 J of heat into the hot reservoir. So running engine (a) to operate the refrigerator with a COP > 2 will result in the transfer of heat from the cold to the hot reservoir without doing any net mechanical work in violation of the second law.

29* • A heat pump delivers 20 kW to heat a house. The outside temperature is −10°C and the inside temperature of the hot-air supply for the heating fan is 40°C. (a) What is the coefficient of performance of a Carnot heat pump operating between these temperatures? (*b*) What must be the minimum power of the engine needed to run the heat pump? (*c*) If the COP of the heat pump is 60% of the efficiency of an ideal pump, what must the minimum power of the engine be?

(*a*) COP = $T_c/\Delta T$ (see Problem 21)   COP = 263/50 = 5.26

(*b*) Use Equ. 20-10; $P = W/t$ $P = [20/(1 + 5.26)]$ kW = 3.19 kW

(*c*) $P' = P/0.6$ $P' = 3.19/0.6$ kW = 5.32 kW

33* ••   On a humid day, water vapor condenses on a cold surface. During condensation, the entropy of the water

(*a*) increases.

(*b*) remains constant.

(*c*) decreases.

(*d*) may decrease or remain unchanged.

(*c*)

37* •   What is the change in entropy of 1.0 kg of water when it changes to steam at 100°C and a pressure of 1 atm?

$\Delta S = \Delta Q/T$ $\Delta S = 2257/373$ kJ/K = 6.05 kJ/K

41* ••   A system absorbs 300 J from a reservoir at 300 K and 200 J from a reservoir at 400 K. It then returns to its original state, doing 100 J of work and rejecting 400 J of heat to a reservoir at a temperature *T*. (*a*) What is the entropy change of the system for the complete cycle? (*b*) If the cycle is reversible, what is the temperature *T*?

(*a*) *S* is a state function of the system. ΔS for complete cycle = 0.

(*b*) $\Delta S = Q_1/T_1 + Q_2/T_2 + Q_3/T_3 = 0$; solve for T_3 1 J/K + 0.5 J/K − (400 J)/T_3 = 0; $T_3 = T = 267$ K

45* ••   A 1-kg block of copper at 100°C is placed in a calorimeter of negligible heat capacity containing 4 L of water at 0°C. Find the entropy change of (*a*) the copper block, (*b*) the water, and (*c*) the universe.

(*a*) Use the calorimetry equation to find the final temperature $1 \times 0.386(100 - t) = 4 \times 4.184(t - 0); t = 2.26°C$
= 275.4 K

Find $\Delta S_{Cu} = m_{Cu}c_{Cu} \ln(T_f/T_i)$ $\Delta S_{Cu} = 0.386 \ln(275.4/373)$ kJ/K = −117 J/K

(*b*) $\Delta S_w = m_w c_w \ln(T_f/T_i)$ $\Delta S_w = 4 \times 4.184 \ln(275.4/273.15)$ kJ/K = 137 J/K

(*c*) $\Delta S_u = \Delta S_{Cu} + \Delta S_w$ $\Delta S_u = 20$ J/K; $\Delta S_u > 0$, the process is irreversible

---

49* ••   If 500 J of heat is conducted from a reservoir at 400 K to one at 300 K, (*a*) what is the change in entropy of the universe, and (*b*) how much of the 500 J of heat conducted could have been converted into work using a cold reservoir at 300 K?

(*a*) $\Delta S_u = \Delta S_h + \Delta S_c = -Q/T_h + Q/T_c$ $\Delta S_u = 500(1/300 - 1/400) = 0.417$ J/K

(*b*) $\varepsilon_{max} = 1 - T_c/T_h$; $W = \varepsilon Q_h$ $\varepsilon_{max} = 0.25$; $W = 0.25 \times 500 = 125$ J

53* ••   An ideal gas is taken reversibly from an initial state P_i, V_i, T_i to the final state P_f, V_f, T_f. Two possible paths are (A) an isothermal expansion followed by an adiabatic compression, and (B) an adiabatic compression followed by an isothermal expansion. For these two paths,

(*a*) $\Delta U_A > \Delta U_B$.

(b) $\Delta S_A > \Delta S_B$.

(c) $\Delta S_A < \Delta S_B$.

(d) none of the above is correct.

(d)

57* ••   Sketch an SV diagram of the Carnot cycle.

Referring to Figure 20-8, process 1-2 is an isothermal expansion. In this process heat is added to the system and the entropy and volume increase. Process 2-3 is adiabatic, so S is constant as V increases. Process 3-4 is an isothermal compression in which S decreases and V also decreases. Finally, process 4-1 is adiabatic, i.e., isentropic, and S is constant while V decreases. The cycle is shown in the adjacent SV diagram.

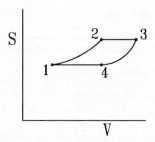

61* •   Which has a greater effect on increasing the efficiency of a Carnot engine, a 5-K increase in the temperature of the hot reservoir or a 5-K decrease in the temperature of the cold reservoir?

Let ΔT be the change in temperature and $\varepsilon = (T_h - T_c)/T_h$ be the initial efficiency. If T_h is increased by ΔT, ε', the new efficiency is $\varepsilon' = (T_h + \Delta T - T_c)/(T_h + \Delta T)$. If T_c is reduced by ΔT, the efficiency is then $\varepsilon'' = (T_h - T_c + \Delta T)/T_h$. The ratio $\varepsilon''/\varepsilon' = T_h/(T_h + \Delta T) > 1$. Therefore, a reduction in the temperature of the cold reservoir by ΔT increases the efficiency more than an equal increase in the temperature of the hot reservoir.

65* ••   The system represented in Figure 20-17 (Problem 54) is 1 mol of an ideal monatomic gas. The temperatures at points A and B are 300 and 750 K, respectively. What is the thermodynamic efficiency of the cyclic process ABCDA?

The cycle is the Carnot cycle. $\varepsilon = 1 - T_c/T_h$ $\varepsilon = 1 - 300/750 = 0.6 = 60\%$

69* ••   A heat engine that does the work of blowing up a balloon at a pressure of 1 atm extracts 4 kJ from a hot reservoir at 120°C. The volume of the balloon increases by 4 L, and heat is exhausted to a cold reservoir at a temperature T_c. If the efficiency of the heat engine is 50% of the efficiency of a Carnot engine working between the same reservoirs, find the temperature T_c.

1. Find W $W = 4$ atm·L $= 0.404$ kJ

2. $\varepsilon = W/Q_h$; $\varepsilon_C = 2\varepsilon = 1 - T_c/T_h$ $\varepsilon_C = 2 \times 0.404/4 = 0.202 = 1 - T_c/393$; $T_c = 313.6$ K

3. $t_c = T_c - 273.15$ $t_c = 40.5°$C

73* ••   Two moles of a diatomic gas are carried through the cycle ABCDA shown in the PV diagram in Figure 20-21. The segment AB represents an isothermal expansion, the segment BC an adiabatic expansion. The pressure and temperature at A are 5 atm and 600 K. The volume at B is twice that at A. The pressure at D is 1atm. (a) What is the pressure at B? (b) What is the temperature at C? (c) Find the work done by the gas in one cycle and the thermodynamic efficiency of this cycle.

(a) 1. $P_B = P_A(V_A/V_B)$ $P_B = 5/2$ atm $= 2.5$ atm $= 252.5$ kPa;

    2. Find $V_B = nRT_B/P_B$ $V_B = 2 \times 8.314 \times 600/2.525 \times 10^5$ m^3 $= 39.5$ L

(b) 1. Find $V_C = V_B(P_B/P_C)^{1/\gamma}$; $\gamma = 1.4$ $V_C = 39.5 \times 2.5^{0.714}$ L $= 76$ L

    2. $T_C = T_B(P_C V_C/P_B V_B)$ $T_C = 600(76/98.75)$ K $= 462$ K

(c) 1. $W_{A-B} = nRT_A \ln(V_B/V_A)$

　　　2. $W_{B-C} = -nc_v\Delta T$

　　　3. $W_{C-D} = P_C(V_D - V_C)$; $W_{D-A} = 0$

　　　4. $W = W_{A-B} + W_{B-C} + W_{C-D} + W_{D-A}$

　　　5. $Q_{D-A} = nc_v(T_A - T_D)$; $T_D = T_A/5$

　　　6. $Q_{in} = Q_{A-B} + Q_{D-A}$; $Q_{A-B} = W_{A-B}$; $\varepsilon = W/Q_{in}$

$W_{A-B} = 2 \times 8.314 \times 600 \times \ln(2)$ J $= 6.915$ kJ

$W_{B-C} = 2 \times 2.5 \times 8.314 \times 138$ J $= 5.74$ kJ

$W_{C-D} = 1(19.75 - 76)$ atm·L $= -5.68$ kJ; $W_{D-A} = 0$

$W = 6.975$ kJ

$Q_{D-A} = 2 \times 2.5 \times 8.314 \times 480$ J $= 20$ kJ

$\varepsilon = 6.975/26.915 = 0.259 = 25.9\%$

77* •• Compare the efficiency of the Stirling cycle (see Figure 20-14) and the Carnot engine operating between the same maximum and minimum temperatures.

The efficiency of the Sterling cycle, ε_S, is given in Problem 15. Using $T_h - T_c = \varepsilon_C T_h$, that expression can be recast in the form

$$\varepsilon_S = \frac{\varepsilon_C}{1 + \dfrac{c_v\varepsilon_C}{R\ln(V_d/V_c)}},$$ where V_c and V_d are the volumes indicated in Figure 20-14. Clearly, $\varepsilon_S < \varepsilon_C$.

81* ••• Suppose that each engine in Figure 20-22 is an ideal reversible heat engine. Engine 1 operates between temperatures T_h and T_m and Engine 2 operates between T_m and T_c, where $T_h > T_m > T_c$. Show that

$$\varepsilon_{net} = 1 - \frac{T_c}{T_h}$$

This means that two reversible heat engines in series are equivalent to one reversible heat engine operating between the hottest and coldest reservoirs.

Since the engines are ideal reversible engines, $\varepsilon_1 = 1 - T_m/T_h$ and $\varepsilon_2 = 1 - T_c/T_m$. Using the expression in Problem 80 one obtains $\varepsilon_{net} = 1 - T_c/T_h$.

CHAPTER **21**

Thermal Properties and Processes

1* • Why does the mercury level first decrease slightly when a thermometer is placed in warm water?

The glass bulb warms and expands first, before the mercury warms and expands.

5* •• (a) Define a coefficient of area expansion. (b) Calculate it for a square and a circle, and show that it is 2 times the coefficient of linear expansion.

(a) $\gamma = \dfrac{\Delta A/A}{\Delta T}$. (b) For a square, $\Delta A = L^2(1 + \alpha\Delta T)^2 - L^2 = L^2(2\alpha\Delta T + \alpha^2\Delta T^2) = A(2\alpha\Delta T + \alpha^2\Delta T^2)$; in the limit $\Delta T \to 0$, $\Delta A/A = 2\alpha\Delta T$, and $\gamma = 2\alpha$. For the circle, proceed in same way except that now $A = \pi R^2$; again, $\gamma = 2\alpha$.

9* •• A container is filled to the brim with 1.4 L of mercury at 20°C. When the temperature of container and mercury is raised to 60°C, 7.5 mL of mercury spill over the brim of the container. Determine the linear expansion coefficient of the container.

1. Express problem statement in terms of V and ΔV $V_{Hg} = V_c = 1.4$ L; $\Delta V_{Hg} - \Delta V_c = 7.5 \times 10^{-3}$ L

2. Apply Equ. 21-4 and solve for $\beta_{Hg} - \beta_c$ $\beta_{Hg} - \beta_c = [7.5 \times 10^{-3}/(1.4 \times 40)]$ K^{-1} = 1.34×10^{-4} K^{-1}

3. Solve for β_c and apply Equ. 21-5 $\beta_c = (1.8 - 1.34) \times 10^{-4}$ K^{-1} = 0.46×10^{-4} K^{-1}

                                     $\alpha = 15 \times 10^{-6}$ K^{-1}

13* •• A car has a 60-L steel gas tank filled to the top with gasoline when the temperature is 10°C. The coefficient of volume expansion of gasoline is $\beta = 0.900 \times 10^{-3}$ K^{-1}. Taking the expansion of the steel tank into account, how much gasoline spills out of the tank when the car is parked in the sun and its temperature rises to 25°C?

Spill = $\Delta V_{gas} - \Delta V_{tank} = V\Delta T(\beta_{gas} - \beta_{tank})$ Spill = $(60)(15)(9 \times 10^{-4} - 3 \times 11 \times 10^{-6})$ L = 0.78 L

17* ••• A steel tube has an outside diameter of 3.000 cm at room temperature (20°C). A brass tube has an inside diameter of 2.997 cm at the same temperature. To what temperature must the ends of the tubes be heated if the steel tube is to be inserted into the brass tube?

$r_s(1 + \alpha_s\Delta T) = r_b(1 + \alpha_b\Delta T)$; $\Delta T = \dfrac{r_s - r_b}{\alpha_b r_b - \alpha_s r_s}$ $\Delta T = \dfrac{3.000 - 2.997}{19 \times 10^{-6} \times 2.997 - 11 \times 10^{-6} \times 3.000}$ K = 125 K

$T = T_0 + \Delta T$ $T = (20 + 125)°C = 145°C$

21* ••   The phase diagram in Figure 21-14 can be interpreted to yield information on how the boiling and melting points of water change with altitude. (*a*) Explain how this information can be obtained. (*b*) How might this information affect cooking procedures in the mountains?

(*a*) With increasing altitude *P* decreases; from curve *OF*, *T* of liquid-gas interphase diminishes, so the boiling temperature decreases. Likewise, from curve *OH*, the melting temperature increases with increasing altitude.

(*b*) Boiling at a lower temperature means that the cooking time will have to be increased.

25* ••   The van der Waals constants for helium are $a = 0.03412$ L$^2 \cdot$ atm/ mol^2 and $b = 0.0237$ L/mol. Use these data to find the volume in cubic centimeters occupied by one helium atom and to estimate the radius of the atom.

In Equ. 21-6, b = volume of 1 mol of molecules $(0.0237$ L/mol)(1 mol/6.022×10^{23} atoms)(10^3 cm^3/1 L)

For He, 1 molecule = 1 atom $= 3.94 \times 10^{-23}$ cm^3/atom

$V = (4/3) \pi r^3$; solve for r $r = (3 \times 3.94 \times 10^{-23}/4\pi)^{1/3} = 2.11 \times 10^{-8}$ cm $= 0.211$ nm

---

29* ••   Two metal cubes with 3-cm edges, one copper (Cu) and one aluminum (Al), are arranged as shown in Figure 21-15. Find (*a*) the thermal resistance of each cube, (*b*) the thermal resistance of the two-cube system, (*c*) the thermal current *I*, and (*d*) the temperature at the interface of the two cubes.

(*a*) Use Equ.21-10; substitute numerical values $R_{Cu} = 1/(0.03 \times 401) = 0.0831$ K/W; $R_{Al} = 0.141$ K/W

(*b*) $R = R_{Cu} + R_{Al}$ $R = 0.224$ K/W

(*c*) $I = \Delta T/R$ $I = 80/0.224$ W $= 358$ W

(*d*) $I_{Cu} = I_{Al} = I$; $\Delta T_{Cu} = I_{Cu}R_{Cu}$ $\Delta T_{Cu} = 358 \times 0.0831$ K $= 29.7$ K; $T = 100 - 29.7$
    $= 70.3$°C

33* ••   For a boiler at a power station, heat must be transferred to boiling water at the rate of 3 GW. The boiling water passes through copper pipes having a wall thickness of 4.0 mm and a surface area of 0.12 m^2 per meter length of pipe. Find the total length of pipe (actually there are many pipes in parallel) that must pass through the furnace if the steam temperature is 225°C and the external temperature of the pipes is 600°C.

1. From Equ. 21-7, $A = I\Delta x/k\Delta T$ $A = (3 \times 10^9)(4 \times 10^{-3})/[(401)(375)]$ m$^2 = 79.8$ m^2

2. $L = A/(0.12$ m) $L = 665$ m

37* •   Calculate λ_{max} for a human blackbody radiator, assuming the surface temperature of the skin to be 33°C. Use Equ. 21-21 and substitute numerical values. $\lambda_{max} = (2.898 \times 10^{-3})/(273 + 33)$ m $= 9.47$ μm

---

41* ••   Liquid helium is stored at its boiling point (4.2 K) in a spherical can that is separated by a vacuum space from a surrounding shield that is maintained at the temperature of liquid nitrogen (77 K). If the can is 30 cm in diameter and is blackened on the outside so that it acts as a blackbody, how much helium boils away per hour?

1. $dm/dt = P_{net}/L = e\sigma \pi d^2(T^4 - T_0^4)/L$. Here L is the $dm/dt = (5.67 \times 10^{-8} \times \pi \times 0.3^2 \times 77^4/21 \times 10^3)$ kg/s

latent heat of boiling. $T_0 = 4.2$ K can be $= 2.68 \times 10^{-5}$ kg/s $= 9.66 \times 10^{-2}$ kg/h $= 96.6$ g/h

neglected compared to $T = 77$ K

45* •   The earth loses heat by (*a*) conduction. (*b*) convection. (*c*) radiation. (*d*) all of the above.

(*c*)

49* •   A steel tape is placed around the earth at the equator when the temperature is $0°C$. What will the clearance between the tape and the ground (assumed to be uniform) be if the temperature of the tape rises to $30°C$? Neglect the expansion of the earth.

From Equ. 21-2, $\Delta R = R\alpha\Delta T$

$\Delta R = 6.38 \times 10^6 \times 11 \times 10^{-6} \times 30$ m

$= 2.1 \times 10^3$ m $= 2.1$ km

53* ••   Lou has patented a cooking timer, which he is marketing as "Nature's Way: Taking You Back To Simpler Times." The timer consists of a 28-cm copper rod having a 5.0-cm diameter. Just as the lower end is placed in boiling water, an ice cube is placed on the top of the rod. When the ice melts completely, the cooking time is up. A special ice cube tray makes cubes of various sizes to correspond to the boiling time required. What is the cooking time when a 30-g ice cube at $-5.0°C$ is used?

1. Find t_1, time to raise the ice cube from $-5°C$ to $0°C$.

$\Delta Q = (0.03 \text{ kg})(2.05 \times 10^3 \text{ J/kg·K})(5 \text{ K}) = 307.5 \text{ J}$

$t_1 = \Delta Q/[kA(\Delta T/\Delta x)]$; $\Delta Q = mc \times (5 \text{ K})$

$t_1 = 307.5/[401 \times (\pi \times 25 \times 10^{-4}/4) \times (102.5/0.28)]$ s

$= 1.07$ s

2. t_2, time to melt ice $= mL\Delta x/kA\Delta T$

$t_2 = \left(\dfrac{0.03 \times 333.5 \times 10^3 \times 0.28}{401 \times (25 \times 10^{-4}\pi/4) \times 100} \right)$ s $= 35.6$ s

3. The total time $= t_1 + t_2$

$t_{tot} = 36.7$ s

57* ••   (a) From the definition of β, the coefficient of volume expansion (at constant pressure), show that $\beta = 1/T$ for an ideal gas. (b) The experimentally determined value of β for N_2 gas at $0°C$ is 0.003673 K^{-1}. Compare this value with the theoretical value $\beta = 1/T$, assuming that N_2 is an ideal gas.

(a) For an ideal gas, $V = nRT/P$; $\beta = (1/V)(dV/dT) = (P/nRT)(nR/P) = 1/T$. (b) $1/273 = 0.003663$ is within 0.3 % of the experimental value.

61* ••   A hot-water tank of cylindrical shape has an inside diameter of 0.55 m and inside height of 1.2 m. The tank is enclosed with a 5-cm-thick insulating layer of glass wool whose thermal conductivity is 0.035 W/m·K. The metallic interior and exterior walls of the container have thermal conductivities that are much greater than that of the glass wool. How much power must be supplied to this tank to maintain the water temperature at $75°C$ when the external temperature is $1°C$?

We will do this problem twice. First, we shall disregard the fact that the surrounding insulation is cylindrical. We shall then repeat the problem, proceeding as in Problem 34.

(a) 1. Find the total area

$A_{tot} = [2 \times (\pi/4)(0.55)^2 + \pi \times 0.55 \times 1.2]$ m^2 = 2.55 m^2

2. Use Equ. 21-7

$I = (0.035)(2.55)(74/0.05)$ W = 132 W

(b) 1. Find I through the top and bottom surfaces - I_1

$I_1 = (0.035)[(\pi/2)(0.55)^2](74/0.05)$ W = 24.6 W

2. Find I_c through the cylindrical surface. Consider an element of cylindrical area of length L, radius r, and thickness dr. The heat current is $I_c = -2\pi kLr(dT/dr)$. Thus, $dT = -[I_c/(2\pi kL)]dr/r$. Integrate from T_1 to T_2 and from r_1 to r_2 and solve for the heat current I_c.

$I_c = 2\pi kL(T_1 - T_2)/\ln(r_1/r_2)$.

$I_c = 2\pi(0.035)(1.2)(74)/\ln(0.65/0.55)$ W = 97.4 W

3. Find total heat loss $I = I_1 + I_c$

$I = (24.6 + 97.4)$ W = 122 W

65* ••• A body initially at a temperature T_i cools by convection and radiation in a room where the temperature is T_0. The body obeys Newton's law of cooling, which can be written $dQ/dt = hA(T - T_0)$, where A is the area of the body and h is a constant called the surface coefficient of heat transfer. Show that the temperature T at any time t is given by $T = T_0 + (T_i - T_0)e^{-hAt/mc}$, where m is the mass of the body and c is its specific heat.

1. $dQ = -mcdT$ is heat loss as T diminishes by dT. Thus, $dT = -(1/mc)dQ$ and $dT/dt = -(hA/mc)(T - T_0)$

2. $\displaystyle\int_{T_i}^{T} \frac{dT}{T - T_0} = -\frac{hA}{mc}\int_0^t dt; \quad \ln\left(\frac{T - T_0}{T_i - T_0}\right) = -\frac{hA}{mc}t$, where T_i is the initial temperature.

Take the antilog and solve for T to obtain $T = T_0 + (T_i - T_0)e^{-hAt/mc}$.